Edible Wild Plants

Effective Tips and Tricks to Procuring Nutritious

(A Comprehensive Beginner's Guide to Learn the Realms of Foraging)

William Woods

Published By **George Denver**

William Woods

All Rights Reserved

Edible Wild Plants: Effective Tips and Tricks to Procuring Nutritious (A Comprehensive Beginner's Guide to Learn the Realms of Foraging)

ISBN 978-0-9949563-8-5

No part of this guidebook shall be reproduced in any form without permission in writing from the publisher except in the case of brief quotations embodied in critical articles or reviews.

Legal & Disclaimer

The information contained in this book is not designed to replace or take the place of any form of medicine or professional medical advice. The information in this book has been provided for educational & entertainment purposes only.

The information contained in this book has been compiled from sources deemed reliable, and it is accurate to the best of the Author's knowledge; however, the Author cannot guarantee its accuracy and validity and cannot be held liable for any errors or omissions. Changes are periodically made to this book. You must consult your doctor or get professional medical advice before using any of the suggested remedies, techniques, or information in this book.

Upon using the information contained in this book, you agree to hold harmless the Author from and against any damages, costs, and expenses, including any legal fees potentially resulting from the application of any of the information provided by this guide. This disclaimer applies to any damages or injury caused by the use and application, whether directly or indirectly, of any advice or information presented, whether for breach of contract, tort, negligence, personal injury, criminal intent, or under any other cause of action.

You agree to accept all risks of using the information presented inside this book. You need to consult a professional medical practitioner in order to ensure you are both able and healthy enough to participate in this program.

Table Of Contents

Chapter 1: The Heart Of Foraging 1

Chapter 2: Tools Of The Trade 15

Chapter 3: The Art Of Identification 29

Chapter 4: Foraging Through The Seasons 42

Chapter 5: Safe And Sustainable Harvesting 56

Chapter 6: Edible Landscapes And Habitats 70

Chapter 7: From Field To Plate 84

Chapter 8: Medicinal And Practical Uses 98

Chapter 9: Building Community Around Foraging 110

Chapter 10: Global Foraging 123

Chapter 11: Identifying Edible Wild Plants 132

Chapter 12: Foraging Techniques 141

Chapter 13: Common Edible Wild Plants 159

Chapter 14: Safety Precautions............171

Chapter 1: The Heart Of Foraging

Nature, in her majestic vastness, has prolonged held solutions and solace for souls in is looking for. Through millennia, humankind's deep-seated reference to the land has manifested inside the artwork of foraging—an act that transcends mere survival, accomplishing into geographical regions of ethics, data, and spirituality. While the contemporary age regularly feels disconnected from the wild, foraging stays a poignant exercising, offering pathways to understand our beyond, act ethically in the gift, and join spiritually with the essence of life. At the coronary heart of foraging lies a treasure trove of insights, geared up to be unveiled and cherished.

The Ethical Foundations of Foraging

Foraging is more than only a meandering walk through the woods, plucking plant life at whim. It's an age-antique way of life, imbued with a profound apprehend for the land and its bounties. At the coronary heart of this workout lies a moral foundation, deeply rooted in records, reverence, and sustainability.

Many modern-day societies carry out under a consumerist paradigm, in which meals is a product to be offered, consumed, and discarded. Supermarkets are laden with devices shipped from lots of miles away, regularly grown beneath doubtful environmental situations, and packaged in wasteful substances. We've distanced ourselves from the origins of our sustenance, dropping a connection now not most effective to the earth but additionally to the very act of survival and coexistence.

In stark assessment, foraging reconnects us to the ones forgotten threads. The ethical cloth of foraging is woven with strands of sustainability, mindfulness, and a deep-rooted respect for Mother Earth. When one devices out to forage, they may be not virtually 'taking' from nature. They're getting into a silent agreement, a promise to tread gently, to awesome harvest what is wanted, and to make sure the ongoing flourishing of herbal habitats.

The global spherical us pulses with life, every twig, berry, and leaf gambling a thing in a grand, intricate dance of ecosystems. Foragers understand this. They understand that plucking a more youthful sapling might deprive the place of a destiny tree, or that uprooting a plant can disrupt the touchy balance of a micro-environment. There's an innate sense of responsibility that drives the moves of a true forager, making sure their choices are in harmony with the environment.

One might also surprise, then, how does one extend this ethical compass? How does one decide what is proper from wrong inside the top notch tapestry of nature?

The key lies in training and empathy. To forage ethically is to be continuously reading: about flowers, their existence cycles, their roles inside the surroundings, and their relationships with different organisms. But this facts isn't always in simple terms academic. It's experiential. It's acquired through countless hours looking at nature, expertise its rhythms, and immersing oneself in its wonders.

Beyond knowledge, empathy plays a critical characteristic. To empathize with nature is to look it now not as a aid but as a living, breathing entity. It's to enjoy a feel of kinship with the timber, the birds, the bugs, and to apprehend that our actions reverberate in this shared area we name Earth. This empathy translates into picks that prioritize the well-being of nature over personal benefit.

While the idea would probably sound idyllic, it's furthermore pragmatic. For if we over-harvest, if we disrupt habitats, if we approach foraging with avarice in desire to apprehend, we are not truely harming nature. We're undermining our very very own future functionality to forage, to benefit from nature's bounty. Ethical foraging, then, is as a exquisite deal approximately self-protection as it's miles approximately retaining the surroundings.

Consider, for a 2nd, a sprawling very welltree. Its branches stretch out, offering refuge to countless birds. Its roots dig deep, drawing nourishment from the earth and stabilizing the soil. Acorns drop, imparting sustenance for a plethora of creatures, from squirrels to deer. Now agree with a forager coming near this tree. An ethical forager might not clearly see a functionality meal in the form of acorns. They'd see a microcosm of existence, a node in the sizable net of lifestyles. Their actions, whether or not or now not they decide to collect acorns or now not, would possibly stem from this deep appreciation and understanding.

It's truly really worth noting that ethical foraging is not a static idea. It evolves, fashioned via changing environments, societal values, and our developing facts of ecology. What remains constant, however, is the underlying ethos: a willpower to harmony, stability, and appreciate.

So, foraging, while approached with an ethical coronary coronary heart, isn't always in fact an act of gathering meals. It will become a profound expression of our courting with the earth. It's a mild whisper, a silent pledge, echoing thru the woods, the meadows, and the marshes: "I am proper here, not as a conqueror, but as a mother or father. I shall take, however I shall moreover supply. For in your well-being lies my personal."

Foraging, then, isn't pretty lots survival. It's approximately coexistence. It's about know-how that within the grand tapestry of life, we're however a single thread, intertwined with limitless others. And it's miles this recognition, this profound humility, that office work the ethical bedrock of foraging.

A Historical Journey: From Nomads to Modern Foragers

At the dawning of humanity, at the same time as the first testimonies were whispered beneath starlit skies and civilizations were however unborn dreams, our ancestors have been foragers. This ancient workout, as crucial because the breath we take, tells a story that spans epochs, tracing the evolution of humanity from its humble beginnings to the complexities of our cutting-edge-day age.

Long in advance than concrete jungles and bustling metropolises, incredible stretches of wild land stretched as a ways as the attention ought to see. In those primordial landscapes, nomadic tribes wandered, counting on their intimate expertise of the earth to preserve them. These were not mere wanderers; they had been the unique foragers, the pioneers who navigated thru nature's rhythm, extracting sustenance from the land, the water, and the sky.

The life of early nomads turn out to be shaped via the seasons. Spring brought forth a burst of greenery, with glowing shoots, berries, and fruits becoming ripe for the picking. Summer supplied an abundance of leafy plants and a greater variety of cease end result. Autumn heralded the harvest of nuts and seeds, on the equal time as wintry climate, with its stark landscape, shifted the focus to searching and fishing.

But it wasn't simply about gathering food. These nomads have been nature's scholars. They observed the conduct of animals, studying from their behaviors. They noticed how extremely good vegetation grew near water assets, how particular berries attracted unique birds, and the way the moon's degrees brought about the tides. This wealth of records wasn't documented in books or scrolls but exceeded down via memories, songs, and shared testimonies.

As millennia handed, the story started out to shift. With the advent of agriculture round 10,000 years in the beyond, many nomadic tribes transitioned to settled groups. Farming allowed for meals surplus, fundamental to larger populations and the transport of early civilizations. While this shift introduced severa improvements, something intangible have become left in the back of on the ones ancient trails: the profound connection among people and the wild.

Yet, the coronary coronary coronary heart of foraging in no manner actually vanished. Even as empires rose and fell, there remained wallet of people for whom foraging turn out to be no longer handiest a remnant of a bygone generation however a living, respiration way of lifestyles. In numerous cultures, from the Native Americans to the tribes of Africa, foraging persisted to be an important part of their manner of life, a thread connecting them to their ancestors and the land they known as domestic.

Fast earlier to the industrial age, a period marked thru fast technological improvements and urbanization. As towns extended and forests diminished, the art work of foraging placed itself overshadowed by the enchantment of present day conveniences. Supermarkets stocked with goods from across the place changed nearby markets. Nature's bounty, as soon as reliable, become now frequently omitted or, worse, seen as primitive.

But as with all matters, the pendulum of time swings in cycles. In modern-day decades, there may be been a renaissance, a renewed hobby in foraging. Modern foragers are a various institution, a tapestry of people sure via using a shared love for nature and a choice to reconnect with the earth.

Driven through using severa motivations — be it sustainability, a quest for organic produce, or simply the delight of immersion in nature — the ones people are rediscovering the knowledge in their ancestors. But they're also innovating, merging historic practices with modern-day knowledge. They're documenting their findings, sharing them on social media, and forming organizations in which novices and specialists alike can research and expand.

Yet, at the same time as contemporary foragers encompass the digital age, on the heart of their adventure lies some component undying: a deep-seated admire for the land. They tread gently, information that nature isn't an inexhaustible useful aid however a touchy balance of interdependent systems.

In final, the adventure from nomads to trendy foragers isn't always only a story of humanity's evolution however a testament to our enduring bond with nature. It speaks of our innate desire to be a part of something large, to discover our location within the complex dance of lifestyles.

In the silence of the woods, amidst the rustling leaves and the slight hum of existence, echoes the whisper of our ancestors, reminding us of who we as soon as had been and, in all likelihood, guiding us in the path of who we might grow to be. Foraging, then, is not handiest a exercise. It's a bridge, spanning in some unspecified time in the future of time, linking us to our beyond and paving the manner for a destiny wherein humanity and nature coexist in harmonious rhythm.

The Spiritual Connection: Nature and Self

In the midst of the never-ending cacophony of current life, wherein cities by no means sleep and screens glow grade by grade, many souls find themselves craving for some thing deeper. A connection, profound and historical, beckons from the coronary coronary heart of the barren region. This siren song is the whisper of the timber, the murmur of rivers, and the gentle caress of the wind in opposition to one's face. It speaks of a bond, an intertwining of nature and self, that many have forgotten but which, with a touch intention, may be rediscovered.

To forage is to commune with the earth. It's a deliberate act of undertaking out, of in search of, of finding. With every step into the wild, the forager retraces the footprints of our ancestors, rekindling a relationship that predates the confines of civilization. But this relationship is not simply ancient or physical; it's far deeply religious.

Consider the immediately of discovery, whilst a forager's eyes alight upon a coveted mushroom or berry. In that immediate, there may be an first-rate enjoy of gratitude, a interest that the earth, in its boundless generosity, has supplied up a present. This isn't always the indifferent, transactional alternate we are aware about in supermarkets. It's personal. It's intimate. It looks like a shared mystery some of the forager and the land.

Such moments, fleeting as they might be, evoke a profound feel of belonging. In the great tapestry of existence, amid mountains and valleys, forests and deserts, the individual well-knownshows their area. The hobby dawns that we are not separate entities certainly occupying space however important factors of a larger, mind-blowing entire.

Some non secular traditions talk of the concept of oneness, the idea that all of advent is interconnected. Foraging, in its essence, embodies this philosophy. When one stands amidst the bushes, paying attention to the heart beat of the area, differences blur. The separation between self and nature begins to dissolve, and in its region arises a sense of harmony.

Sam, a present day-day-day forager from Oregon, as soon as stated an revel in that captures this sentiment. While out on simply certainly one of his normal tours, he stumbled upon a clearing bathed in the mild glow of the placing sun. The air become thick with the fragrance of pine and earth. As he knelt to investigate a few herbs, he changed into overcome with a sensation that he grow to be not by myself. Not inside the eerie enjoy, but in a comforting, encompassing way. He felt as even though the wooded region itself become acknowledging his presence, welcoming him. In that second, Sam wasn't simplest a tourist; he grow to be a player within the grand symphony of lifestyles.

Chapter 2: Tools Of The Trade

Venturing into the woods and meadows, the forager's adventure is as masses approximately the device they devise due to the fact the treasures they are trying to find. These gadgets, whether homemade or preserve-offered, are the bridge amongst nature's secrets and our human hobby. They encompass the heartbeats of endless expeditions and reminiscences of discovery, becoming witnesses to the ever-evolving relationship amongst mankind and the wild. But like every seasoned vacationer can attest, a device is fine as pinnacle because of the reality the care it gets. By expertise the essence of our device, we not most effective make sure our protection however furthermore honor the age-antique life-style of foraging.

Essential Equipment for Every Forager

Stepping into the world of foraging can often revel in like coming into an arcane realm, wherein age-vintage traditions meet the coronary heart beat of Mother Nature. As you embark in this mesmerizing adventure, just like a maestro getting ready for a symphony or a painter installing his easel, having the proper system turns into paramount. The magic of foraging does no longer in fact lie within the harvest; it is living inside the communion between nature and forager, and the tools play a pivotal function in shaping this connection.

To start with, bear in mind the not unusual-or-lawn basket. Baskets are not virtually the packing containers of your wild harvest; they will be the silent witnesses to your journeys, the keepers of recollections, and often, the bearers of tales that span generations. Woven with care, a top notch basket allows for ok airflow, ensuring that your accumulated devices live sparkling. A sturdy control ought to make the difference among a comfortable haul and a strained wrist. But extra than its functionality, a basket is a photo. It's a willpower to nature's bounty and an embodiment of the forager's ethos — to take most effective what is needed, generally leaving within the again of extra than what's taken.

A notable knife is a few different important tool in a forager's repertoire. With it, viable deftly reduce stems with out causing useless harm to the plant or, in a few times, to extract roots with precision. When selecting a knife, search for one with a blade forged from stainless steel, making sure it stays immune to corrosion. The address want to experience ergonomic, melding together along with your hand like an extension of your very being. Remember, this knife may be your partner in the course of endless forays into the wild, and as such, it need to enjoy as familiar and snug as an vintage friend's embody.

Good footwear, frequently neglected, can be the difference between a profitable foraging experience and an uncomfortable ordeal. Traversing thru forests, meadows, and wetlands, a forager's toes are his maximum trusted allies. The proper pair of footwear will provide the crucial grip on slippery terrains, protect towards sharp gadgets, and offer the considered necessary help for hours of walking. Waterproof materials and breathable fabric can increase your foraging experience, ensuring that even nature's unpredictable mood swings – be it a surprising downpour or an sudden glide crossing – might not dampen your spirits.

While many gadget increase the act of foraging, none is as treasured as knowledge. It might in all likelihood seem intangible, but an first-rate discipline guide is as important as any physical tool. In its pages lie the information of the a long time, guiding you on what to are looking for and what to avoid. A nicely-curated manual will not simplest have special descriptions however also complex illustrations, helping you in distinguishing among appearance-alike vegetation, some of which might be dangerous. While the virtual age has added forth a plethora of apps and on line assets, the tactile enjoy of flipping through a manual, with notes scribbled in margins and bookmarks marking favored pages, brings a feel of nostalgia and grounding that technology regularly fails to duplicate.

Lastly, and possibly most significantly, is the coronary heart of a forager. Yes, it's far now not a device you should buy or craft, but it's miles an important device even though. As you journey thru nature's vastness, permit your coronary heart be your compass, guiding you with a profound recognize for the environment, an insatiable interest to research, and an unwavering dedication to tread lightly and responsibly.

In the arena of foraging, as you tread softly within the global's bosom, collecting nature's treasures, it is critical to do not forget that the tool you deliver are more than mere gadgets. They are extensions of your motive, reflections of your ethos, and silent companions to your dance with nature. Choose them with care, wield them with apprehend, and they'll, in flip, enhance your foraging expeditions, making each time out a memorable symphony of senses and reminiscences.

Crafting Your Own Foraging Kit: A DIY Guide

Venturing into the woods with out the proper gadget can be likened to a sailor navigating the superb oceans without a compass. While modern markets are flooded with enterprise foraging tool, there's an ineffable attraction and intimate connection strong whilst one crafts their very very personal package deal deal. When we create with our private fingers, we're now not simply making equipment; we're weaving our private narratives into the very cloth of the foraging enjoy.

Imagine the hands of our ancestors, raw and earth-stained, molding and fashioning tools for their each day communion with nature. Our journey right here is to emulate that intimate relationship. By crafting our very very very own foraging device, we resonate with the practices of people who walked the earth in advance than us, in detail interwoven with the land they loved.

At the heart of any forager's series is the basket, a vessel that holds nature's treasures. Instead of purchasing a machine-made discipline, there can be some thing deeply attractive about developing your very personal. With supple, lengthy vines or bendy branches, particularly from willow or hazel bushes, you could weave your private story. Beginning at the bottom, weave in a round sample, allowing the rhythm of the craft to resonate along with your heartbeat, making prepared you for the bounty it'll soon keep.

A knife is important, serving not simply as a device, however as an extension of the forager. Crafting your very very own blade may additionally sound daunting, but the rewards are giant. A piece of immoderate-carbon metallic, probably repurposed from an antique file or a fragment of a noticed, can be long-hooked up and sharpened. The deal with, carved from seasoned hardwood like cherry or oak, becomes more than only a grip; it is a testament to the timber which have stood sentinel for years. Binding the two with leather strips or sinew, the knife will become a aggregate of characteristic and sentiment.

Small famous from the wooded vicinity floor, like seeds, berries, or valuable herbs, frequently require a sensitive contact and a stable home. Hand-stitched pouches made from natural cloth like cotton or linen serve this purpose pretty. Infusing these pouches with herbal dyes derived from berries or leaves gives a non-public touch, making every pouch a canvas of your foraging recollections.

While many turn to virtual resources for knowledge, there may be some element to be stated approximately keeping a personal problem manual. A easy pocket book may be converted right into a dwelling report of your foraging escapades. Sketches of flora, non-public observations, or even an occasional pressed leaf make it not only a manual, however a mag of your evolving dating with the wild.

Finally, no foraging tour is entire without the crucial canteen. Eschewing plastic, one ought to probably flip to crafting a very unique location from gourds. Once hollowed and dried, the gourd may be sealed from inside the usage of beeswax, offering you with a useful, and notably non-public, hydration answer.

The act of making your very very own foraging package deal is an adventure in itself. It isn't always sincerely approximately the gear but the memories they carry about about, the recollections they house, and the deep connection they forge between the forager and the land. With each crafted piece, you aren't sincerely making ready for a foraging day journey; you're honoring a undying way of life, echoing the heartfelt dances of ancestors prolonged lengthy past but whispering their knowledge in each crafted curve and detail.

Maintenance and Care: Ensuring Longevity and Safety

The shimmering morning dew, the gentle whispers of the wind, and the crispness of fallen leaves; those are the each day encounters of a forager. Just as the ones wonders of nature are to be cherished, so too must be the device that accompany one on such sojourns. Tools, lovingly crafted or maybe save-bought, name for our interest and care. Beyond most effective a count of durability, the nicely-being of our device speaks to our dedication to the art work of foraging and the honor we pay to nature.

Let's begin with the coronary coronary heart of our series – the forager's basket. Over time, its weaves can also come free, its fibers can also fray, but with the right contact, it is able to be saved sturdy and equipped. If manufactured from natural fibers like willow or hazel, the basket benefits from an occasional spritz of water to prevent it from turning into brittle. After each day revel in, it is properly well worth taking a moment to do away with any residual dust or particles from it, ensuring that the treasures of day after today do no longer mix with the remnants of nowadays. For the ones who have protected their baskets with fabric, an occasional moderate hand wash can keep it fresh and welcoming.

Our trusted knife, that silent associate on every ride, holds reminiscences of limitless discoveries. To keep it sharp and rust-loose, one can also oil it lightly with a natural lubricant. Flaxseed oil or mineral oil, just a dab, massaged onto the blade now not most effective protects it but moreover continues it glowing and equipped. The address, in particular if timber, also can be sometimes oiled to keep its luster and save you it from splitting. After every use, cleaning the blade with a moist fabric after which drying it ensures that it stays in most dependable circumstance, ever equipped to help in our foraging endeavors.

Those beautiful hand-stitched pouches that cradle our smaller unearths need their proportion of affection too. Depending at the material used, maximum can be gently hand-washed with moderate soaps. Drying them in the color, rather than beneath the tough solar, helps keep their coloration and power. For those that have been clearly dyed, a pinch of salt within the wash water can assist set the coloration, making sure that the colors stay vibrant and reflective of the adventures they have been a part of.

Your private region manual, full of sketches, notes, and recollections, on the same time as sturdy, blessings from being saved in a dry region. Moisture is the enemy of paper, and a small sachet of silica gel can assist keep undesirable dampness at bay. If you've had the misfortune of having it moist, lightly separate the pages and allow them to air dry to prevent them from sticking together. It's now not quite lots preserving a e-book, however a legacy of your foraging adventure.

As for our gourd canteen, nature's very very own hydration solution, it requires minimum but important care. Emptying it after each use and allowing it to air dry prevents mold or bacteria from taking root. The out of doors, particularly if embellished or carved, can be polished with a hint of beeswax to maintain it sparkling and to act as a defensive barrier.

Taking care of our system goes beyond simply safety. It's a meditative exercise, a way to mirror on our expeditions, and a way to reveal gratitude for the reminiscences cast. Each cleaning stroke, each oiling movement, is a tribute to the land and its bounties. When we take care of our equipment, we are not surely making sure their sturdiness, however furthermore recommitting ourselves to the craft and ethos of foraging. It will become a cyclic dance – the land nurtures us, we nurture our tools, and our device, in flip, assist us further encompass the land.

As we finish this deep dive into the care and protection of our equipment, permit's carry ahead a philosophy: Every device includes a story, a memory, a piece of our coronary coronary heart. They're now not simply inanimate gadgets however extensions of our passion. In making sure their durability and protection, we honor our craft, we admire nature, and we re-light the hearth that drives us to forage, explore, and find out.

Chapter 3: The Art Of Identification

The paintings of foraging is as undying because the rhythm of nature itself, a dance of discovery that has resonated via the while. But, because the arena advanced, so did our manner of facts it. No longer is the forager genuinely reliant on tactile recollections or ancestral knowledge handed down through whispered testimonies. Today, this historic dance is enriched by manner of the use of the insights of technological know-how, the marvels of technology, and the collective information of a related global. From understanding the complex styles of flowers to embracing the symphony of senses and harnessing the electricity of virtual wonders, the current-day forager is each a scholar and a teacher, a seeker and a storyteller.

Understanding Plant Structures and Patterns

Nature, in her countless expertise, has sculpted a myriad of botanical wonders, each with its very very own story, each making a song its very own track. The first step to interpreting these whispered testimonies is via manner of greedy the diffused artwork of expertise plant systems and patterns. It's akin to analyzing a new language, in which every leaf, stem, and root has its non-public dialect, a silent lexicon equipped to be understood.

Imagine status in a super library, with books written in a script ordinary to you. This is the state of affairs for many who first project into the arena of foraging. Nature's bookshelf is replete with numerous volumes, every plant a brilliant tome. And to comprehend their memories, we need to first observe their language.

Rooted inside the floor, plant life are eternal storytellers. A mere have a look at a leaf can speak of its lineage, its struggles, and its goals. The serrated edges may additionally suggest a statistics of defensive in the direction of herbivores, at the same time as its smooth ground could probably mirror a existence spent shooting the glistening morning solar. But leaves are clearly the beginning. The stem, often omitted, is the plant's highway, a conduit of existence, ferrying belongings and tales from root to tip. Its thickness, texture, and resilience can frequently trace at the plant's age and its battles toward the factors.

Flowers, however, are the poets of the plant worldwide. With their shades and structures, they weave recollections of affection and attraction, luring pollinators with guarantees of nectar and candy fragrances. The range of petals, their affiliation, and even the intricacies in their reproductive elements can be deciphered to categorise a plant and apprehend its place inside the botanical hierarchy. Not just mere aesthetics, they maintain keys to evolutionary reminiscences and organic techniques that have evolved over eons.

Then there are the roots, hidden mysteries buried within the soil. They are similar to the heartbeats of flowers, pulsating with existence, carrying out out into the earth, grasping, eating, and anchoring. Roots speak of staying energy and tenacity. Their unfold and intensity can talk of the plant's age, its need for assets, or perhaps its dating with the microbial groups with which it shares its subterranean home.

The artwork of identity, therefore, becomes a dance of statement and deduction. The patterns seen in a single plant regularly mirror those in each different. The spiral affiliation of leaves on a stem might also appear precise at the start, but quick you could recognize it's miles a common technique many vegetation hire, a sample referred to as 'phyllotaxis.' Such repeating systems and styles become acquainted buddies to the observant forager. They act as signposts, guiding us deeper into the plant's narrative.

A surprising satisfactory pal on this journey of knowledge is the very environment wherein the ones plants thrive. Wetlands, as an instance, nurture flora with broader leaves, a format that lets in them to maximise photosynthesis in regions with decrease mild.

Rocky terrains, but, are havens for succulents and plants with deeper root structures, developed to tap into the hidden reservoirs of moisture.

But, as with all language, fluency comes now not just from theoretical expertise but moreover from immersion. The more time one spends amidst nature, lightly touching the leaves, examining the vegetation, and watching the complicated styles, the extra intuitive this language becomes.

It's crucial to technique this mastering curve with humility. There might be instances of wrong identification, however each misstep is but a lesson, a bankruptcy in our ongoing adventure of botanical facts. With time and staying energy, the as quickly as-cryptic global of plant life starts offevolved offevolved to unveil itself, sharing with us recollections as vintage as time, whispered via leaves, sung via vegetation, and penned thru roots within the very fabric of the earth.

Color, Smell, and Touch: Engaging All Senses

In a verdant meadow, nature plays its symphony, a cacophony of points of interest, scents, and textures. While the visible grandeur of flowers often dominates our interest, actual connection and knowledge come from delving deeper, from letting every revel in have interaction and communicate with the arena round. It's an intimate waltz, in which color, odor, and make contact with end up the steps, guiding us via the dance of identity.

The vibrancy of colours inside the botanical realm is not handiest a frivolous show. It's a coded message, a beacon that has superior over millennia, catering to the eyes of those it seeks to draw or warn. For instance, the deep purples and blues of many flowers aren't mere spectacles for human admirers. They resonate with bees, who understand those solar shades with extraordinary readability, making sure a mutual alternate – the promise of nectar for the guarantee of pollination. Similarly, the fiery reds and oranges, regularly an insignia of threat within the animal kingdom, serve as a sign in flora too, caution of capability toxicity.

However, to rely totally on sight might be to miss out on the rich tapestries woven with the aid of other senses. Take a second to hold in mind the heady aroma of a blooming jasmine or the pungent scent of beaten wild garlic. Scents within the plant international are profound storytellers. A go along with the drift of perfume can whisper testimonies of nocturnal pollinators, of moths drawn to the intoxicating scent of a moonflower. The acrid aroma of certain foliage is nature's very very own repellent, keeping off herbivores and making sure the plant's survival. It's fascinating to have a look at how even the absence of a sturdy fragrance, especially in some visually appealing vegetation, speaks of a reliance on wind in location of creatures for pollination.

Yet, the journey of sensory exploration could no longer stop proper proper right here. To simply apprehend a plant, one need to include it, quite literally. The tactile experience, the act of feeling, is as enlightening as it's miles grounding. The velvety touch of a sage leaf or the jagged edges of a dandelion exhibits an awful lot approximately a plant's habitat and its evolutionary adventure. Plants that hail from arid terrains, like succulents, frequently have a waxy, thick texture, an model to keep water. In evaluation, the ones from humid rainforests may additionally flaunt huge, paper-thin leaves, designed to maximize daylight absorption and transpiration.

While touch offers right away remarks, it is crucial to maintain with warning. Nature, in all its benevolence, additionally has its arsenal of defenses. Some plants guard their treasures with thorns, spikes, or even microscopic irritants. The infamous stinging nettle, with its hair-like systems full of formic acid, is a testomony to nature's ingenuity in protection. A moderate brush in the direction of it, and one is rewarded with a stinging sensation, a stark reminder of the plant's obstacles.

Drawing on those senses, forging connections through sight, scent, and contact, enriches the foraging enjoy. It shifts the journey from mere identification to a deeper communion. It becomes much less approximately naming and similarly about understanding, tons less about accumulating and additional approximately connecting.

In the midst of this sensory exploration, recollections are forged. The cerulean blue of a particular flower becomes related with a past due summer time midnight, the musky scent of a mushroom remembers a moist morning inside the woods, and the hard texture of tree bark becomes synonymous with the joys of a state-of-the-art discovery.

Foraging, in this moderate, isn't really an act however an experience. It beckons us to step out of our moved fast lives and immerse ourselves in nature's consist of. By attractive all our senses, thru letting sun shades paint our minds, scents transport us, and textures tell their memories, we emerge as greater than mere spectators. We emerge as a part of the narrative, a thread inside the ever-evolving tapestry of lifestyles.

Digital Aids: Apps and Online Communities for Identification

In a world in which the virtual realm often intertwines seamlessly with our tangible life, the historical exercising of foraging, too, has determined its dance accomplice in era. The whispered secrets and techniques of the woods, the memories informed through meadows, and the enigmatic memories of deserts have been given a voice, amplified by means of the usage of the marvel of modern-day gadget. Today, the act of identifying a plant or mushroom is as plenty an stumble upon with nature as it is a rendezvous with pixels and algorithms.

Imagine venturing into the wild, your coronary heart brimming with anticipation, and stumbling upon a plant, its petals reminiscent of the colors of a twilight sky, its perfume a moderate melody playing with the breeze. In days beyond, this will be the begin of a patient quest, flipping through the weathered pages of concern courses or searching out the records of seasoned foragers. Today, the tale unfolds otherwise. With a trifling click on, a photograph comes alive, racing via substantial digital libraries, getting juxtaposed with hundreds and lots of snap shots, and inner moments, the decision and data of the plant stand observed out. Such is the power of identity apps, amalgamating the expertise of infinite botanists and the precision of device learning.

But it is not quite loads brief answers. These system often redecorate the cellphone-toting forager right proper into a citizen scientist. Some apps, in their quest for accuracy, crowdsource information. Every picture uploaded, every identity made, gives to the collective information, refining the device, and making sure that the subsequent purchaser blessings from this shared know-how.

However, apps are just one issue of the digital useful useful resource spectrum. The substantial expanse of the internet is rife with businesses wherein enthusiasts converge, nice with the aid of way in their love for nature. Forums and social media groups dedicated to foraging turn out to be vibrant ecosystems of their very very very own, humming with discussions, pictures, and anecdotes. Here, a novice's question approximately a curious-searching fern gets answered through a forager from midway at a few level in the globe, and a image of an extraordinary mushroom sparks conversations that traverse taxonomy, culinary makes use of, and conservation.

These businesses regularly go beyond the digital realm, number one to workshops, concern journeys, and festivals. The virtual place hence turns into a catalyst, not only for gaining knowledge of but for forging bonds, for the arrival of a tribe positive thru a shared ardour.

Yet, with all its prowess, the virtual realm has its pitfalls. Relying absolutely on an app's verdict can be perilous, mainly whilst foraging for edibles. Machines, regardless of how superior, lack the nuanced statistics, the instinctual revel in that comes from years of experience. A misidentification, specially within the global of wild mushrooms, ought to advocate the distinction amongst a lovely meal and a systematic emergency.

Thus, on the same time as these device are useful, they need to be used judiciously. The outstanding technique often lies in synergy, in letting technology be the aide, not the grasp. Using apps to slender down possibilities, observed through validation via depended on literature or seasoned foragers, guarantees that the marvel of modern device is harnessed without compromising protection.

Chapter 4: Foraging Through The Seasons

The herbal global is a symphony of ever-converting melodies, with each season imparting its precise rhythm and bounty. From the colorful emergence of spring to the reflective embody of autumn, Mother Nature weaves a gastronomic narrative that beckons each forager. As the calendar pages turn, the landscapes metamorphose, unlocking flavors and smells that outline the essence of time. Embarking on this journey, we traverse via the enough mosaics of each season, delving deep into the treasures they tenderly provide to the discerning forager.

Spring's Bounty: Fresh Greens and Early Berries

Spring, the harbinger of renewal, breathes life into a dormant global. As winter's frosty encompass loosens, the earth, like a seasoned artist, crafts a tableau of bright hues and tantalizing scents. It's for the duration of this transformative season that foragers, with hearts entire of anticipation, set forth to gather the primary rewards of nature's cycle.

One of the earliest services to greet the discerning eye of a forager is the sensitive unfurling of easy greens. These aren't your preferred grocery save sorts but wild wonders that maintain memories of survival and resilience. Take the sprightly wooden sorrel, for instance. With its coronary coronary coronary heart-common leaves and tart flavor, it's not incredible a clean deal with but furthermore a lesson in nature's capability to thrive in diverse conditions.

Nestled alongside those colorful greens are the early bloomers — the berries. Their look isn't some thing quick of theatrical. After months of the earth's quiet contemplation, the unexpected burst of berries, like dollops of vivid paint on a muted canvas, is a sight that stirs the soul. The wild strawberries, smaller than their cultivated contrary numbers but complete of an unrivaled intensity of flavor, are a testomony to the adage that the extraordinary matters frequently are available in small applications.

As one wades deeper into the woods, the blueberries begin to make their presence recognized. These berries, with their deep indigo hue, are nature's jewels, sparkling beneath the slight spring sun. Their tangy sweetness is a far-awaited reward for the patient forager. In the moments when one savors the ones berries, there exists a silent communion among guy and nature, a gratitude for the offers that spring bestows.

But the bounty of spring isn't always quite plenty the palate; it's far an enjoy that engages all of the senses. The moderate rustle of the newly sprouted grass underfoot, the symphony of birds celebrating the season's arrival, and the mild caress of the cool breeze carrying the heady scent of blossoms — a number of those factors integrate to create the magic of spring foraging.

However, with the plethora of alternatives available, it is crucial to technique spring foraging with a sense of respect and facts. Every plant and berry tells a story, and while many are match to be eaten delights, a few convey cautionary recollections. The doppelganger of the wild strawberry, as an instance, is the mock strawberry, a non-poisonous however decidedly bland counterpart. Such subtle differences emphasize the importance of thorough knowledge and cautious statement.

To definitely immerse oneself within the essence of spring foraging, it's useful to now not simply recognition on the harvest however also the experience. Imagine crouching beside a babbling brook, hands gingerly choosing watercress, its peppery flavor a colorful reminder of nature's complexity. Or picture the a laugh of discovering a patch of wild raspberries, their fragile drupelets protecting the splendor of spring's promise.

As the instances of spring wane and the anticipation of summer season's richness grows, the forager's basket turns into a reflected image of nature's generosity. Filled with veggies that nourish and berries that satisfaction, it is a tangible testomony to the magic of the season.

In the give up, spring foraging is greater than most effective a culinary day enjoy; it is a adventure of the soul. Each step taken on the easy earth is a step towards information the complicated internet of lifestyles. Each berry tasted and each leaf picked is a celebration of nature's age-vintage rhythm, a rhythm that whispers recollections of rebirth, resilience, and renewal.

Summer's Richness: Fruits, Flowers, and Seeds

The sultry encompass of summer time cloaks the landscape in a colorful dance of colours and fragrances. As spring's freshness gives way to the heady warmth of these solar-filled months, the earth, in its incredible creativity, offers a myriad of gastronomic marvels. It's sooner or later of this radiant season that foragers, inspired through the generous services of nature, meander thru wild terrains, eyes sparkling with anticipation.

From the verdant canopies above to the grassy meadows beneath, summer season is nature's grand carnival. The timber, not content with definitely the mild whisper in their leaves, now proudly show off their succulent surrender result. Wild cherries preserve close tantalizingly, their deep pink orbs, kissed by way of manner of the solar, beckoning with the promise of sweet nectar. Each chew of these cherries is an explosion of summer time's essence, a taste of the wild, untamed splendor of nature.

While cherries crown the wooden, the meadows and clearings have fun their very personal bounty. The blackberries, with their complicated tapestry of candy and tart, are a forager's delight. Clustering on thorny vines, they will be nature's paradox, training us that every now and then, the maximum precious rewards lie in the lower lower back of disturbing situations. As hands acquire out to build up those berries, there's an unspoken expertise, a silent bond lengthy-hooked up among human and nature, constructed on appreciate and admiration.

But summer season's generosity isn't constrained to surrender end result. As one delves deeper into the heart of the wasteland, a carpet of wildflowers unfurls, every petal and perfume telling memories of seasons lengthy past by using using and those however to go back. Edible blossoms, just like the colorful orange daylilies or the sensitive elderflowers, add no longer simply colour and flavor to the forager's basket, but additionally reminiscences and recollections. These vegetation, at the identical time as sensitive to touch, carry internal them the strong spirit of summer time.

And then there are the seeds. Often unnoticed in preference of their more flamboyant peers, seeds are the unsung heroes of the foraging worldwide. Beneath their unassuming exteriors lie powerhouses of vitamins and flavor. The sunflower, repute tall and regal, gives seeds which can be every a treat and a treasure. Crunchy, nutty, and rich, they embody the strength and staying electricity of the season.

Yet, amidst this bountiful harvest, it is critical to tread with facts and instinct. For example, while the intoxicating aroma of wild roses draws many in the direction of their velvety petals, their comparable-searching counterpart, the canine rose, serves as a reminder that nature, in all its splendor, additionally holds mysteries that require understand and expertise.

The essence of summer time foraging is not surely within the collection however moreover inside the communion. Picture this: a warmth afternoon, the daytime filtering through the dense foliage, casting dappled styles at the floor. Amidst this serenity, a forager sits thru a effervescent circulation, fingers stained with berry juice, a crown of wildflowers resting gently on their brow. In this second of quiet reflected photo, the richness of summer season turns into palpable, now not truely in the flavors of the cease quit result and plants, but inside the profound reference to the area round.

As the prolonged days of summer time start their slow descent into the golden encompass of autumn, the forager's adventure evolves. It's a journey that transcends the mere act of collecting. It's an exploration of the soul, a deep dive into the rhythms of nature and the melodies of the coronary heart.

Summer, in all its opulent grandeur, offers not most effective a ceremonial dinner for the palate however a symphony for the senses. Through the culmination that cling, the flora that bloom, and the seeds that promise new beginnings, it whispers memories of life's ephemeral splendor and the everlasting dance of seasons.

Autumn's Harvest: Nuts, Roots, and Late Bloomers

The canvas of nature, as speedy as painted in the fiery colorations of summer time, step by step transforms as autumn unfurls its golden tapestry. This is a season that captures the imagination, ensnaring the senses with its bloodless breezes, rustling leaves, and the intoxicating aroma of the earth on the point of close eye. As summer season recedes, making way for this time of pondered photo and bounty, foragers discover themselves inside the midst of an superb gastronomic theatre.

Imagine wandering amidst a wooded location cowl, now a mosaic of golds, russets, and ambers. The ground, cushioned through manner of fallen leaves, holds the secrets of autumn's most precious services. Nuts. These tiny pills of strength and taste, housed interior their defensive shells, come to be the coveted prize for plenty a forager. The sturdy oak, in its solemn grandeur, drops acorns, a rich supply of vitamins and a testomony to the cycle of existence. Hazelnuts, with their subtle sweetness, lie hidden amidst the underbrush, looking for those eager enough to find out them.

But nuts aren't the first-rate gem stones autumn has to offer. As timber and shrubs retreat, the floor will become the diploma for some other act of this seasonal play. Roots. These underground treasures, which have absorbed the solar's power in the course of the warmer months, are now ripe for the deciding on. Burdock roots, with their crisp texture and earthy taste, provide a taste quite now not like some thing else. Wild carrots, even though smaller and extra fibrous than their cultivated opposite numbers, p.C. A punch with their concentrated essence of the wild.

Yet, on the equal time as many flowers prepare for rest, a few audacious ones choose out this time to bloom. The late bloomers. Goldenrod, with its towering yellow spires, no longer simplest provides vibrancy to the fall panorama but additionally serves as a haven for beneficial bugs. Its subtly sweet taste is an surprising treat on this season of wealthy and earthy tastes. Another famous character is the passionflower, its tricky petals unfolding in defiance of the encroaching relax, supplying both visible and culinary pleasure.

Autumn, in its silent expertise, teaches the forager staying electricity and understand. Not all roots are suitable for consuming, and not all nuts are ripe for the taking. The black walnut, with its hard exterior and staining juice, is a undertaking to gain, however its wealthy, buttery flesh makes the undertaking worthwhile. However, amidst this abundance, the ghostly white tendrils of the lethal autumn crocus function a stark reminder of nature's dual face. Knowledge, paired with an intuitive know-how of the environment, will become the forager's most depended on brilliant pal.

There's an indescribable magic in foraging within the route of autumn. Perhaps it is the play of daylight, now more golden and slanted, developing enchanting styles at the wooded place floor. Or possibly it's miles the profound stillness, punctuated excellent by using the use of manner of the occasional rustle of a critter getting geared up for wintry weather. Every step taken, each breath inhaled, connects the forager deeper to the land. It's as if the earth, in its final act earlier than the deep freeze, sings a lullaby, inviting all to partake in its beneficiant banquet.

Autumn's essence isn't always in reality in its produce however in its very spirit. It's a time of gratitude, of spotting the fleeting nature of lifestyles, and of celebrating the impermanent beauty that surrounds us. For the forager, every nut collected, every root unearthed, becomes a picture of this profound connection to the natural global.

As the final leaves go along with the flow to the floor and the number one frost kisses the land, the forager's basket brims no longer really with food but with memories Memories of solar-dappled forests, of the fun of discovery, and of the deep, unspoken bond with nature.

The cyclical dance of nature gives greater than without a doubt sustenance; it gives a window into the very soul of the Earth. With each season, we're reminded of the impermanence of life and the everlasting splendor it holds. The memories, the tastes, and the memories cast through foraging are undying. They bridge the gap amongst humanity and the surroundings, reminding us that we are but a thread within the complex internet of lifestyles.

Chapter 5: Safe And Sustainable Harvesting

In the slight encompass of the natural worldwide, in which each entity flourishes in symphonic harmony, the forager seeks not definitely to accumulate, however to emerge as part of the narrative. The desert, with its cascading melodies of life, is not just a area of bounty but also of studying. As each season unfurls its treasures, folks who wander its geographical areas want to understand the gravitas of their steps. The dance of foraging is intimate, stressful recognize for every leaf became, each berry picked. This dance guarantees not definitely the sustenance of the seeker, but the persisted strength of the areas they tread upon.

Recognizing and Avoiding Dangerous Lookalikes

Venturing into the wild coronary coronary heart of nature, the attraction of foraging is undeniably captivating. Yet, hidden amidst the verdant thickets and meadows, lies an tough sport of mimicry in which plant life, harmless and venomous alike, often replicate each other, in search of to deceive the untrained eye. It's a dance of nature, in which distinguishing maximum of the benevolent and the lethal becomes a undergo in mind of vital information, shaping the very basis of stable foraging.

Consider the case of the Morel, a connoisseur's satisfaction, and its doppelgänger, the toxic False Morel. To the green, their resemblance might be uncanny, however keen statement well-known diffused variations. The actual Morel boasts a honeycombed look, at the same time as the False Morel's floor is more odd and warty. A pass over in identification proper proper right here must lead one from the fun of a delicacy to the throes of infection.

Equally misleading is the wild carrot, Daucus carota, which stocks a setting semblance with the deadly poison hemlock, Conium maculatum. A plant that lured the exceptional reality seeker Socrates to his surrender, poison hemlock can be deadly if ingested. Yet, with a discerning gaze, versions emerge. The wild carrot's root emits a nice, carroty aroma, even as the hemlock's is uninviting. Moreover, the stem of the wild carrot is hairy, contrasting the smooth and now and again noticed stem of the poison hemlock.

Such deceits aren't restricted to the world of fungi and vegetation. Berries, those little jewels of the forest, can similarly beguile. While the juicy appeal of untamed strawberries is difficult to face as tons as, they have got a unstable imitator in the guise of the mock strawberry, which, in spite of the fact that now not poisonous, offers a bland taste. Similarly, the secure to devour blueberry has its impersonator inside the toxic "tutsan" berry. The key right here lies in keenly watching the foliage and increase patterns, setting the actual apart from the fraudulent.

The artwork of discerning the ones doppelgängers isn't always pretty masses detailing however knowledge the essence of each species. It's about cultivating a profound connection, nearly intuitive, with the surroundings. When you contact a leaf, experience its texture, its veins; whilst you've got were given a take a look at a mushroom, be aware its gills, its spore colour, its habitat. Such deep engagements get to the bottom of the minute intricacies, which frequently end up the saving grace on this game of lookalikes.

It's crucial to undergo in mind that nature doesn't play via manner of our hints. There's no normal code that a cute plant is stable, or an unpleasant mushroom is toxic. The wild parsnip, with its charming yellow plants, can cause severe burns if its sap touches the pores and skin and is then exposed to sunlight. The harmless-looking autumn skullcap, a mushroom, can result in excessive poisoning if consumed.

Stories are rife approximately eager foragers who've had near encounters, regularly with grave consequences, because of a brief-time period lapse in judgment. The very act of foraging, at the same time as grounding and enlightening, needs understand for the labyrinthine strategies of nature. It's approximately being humble sufficient to famend that in the large encyclopedia of the wild, our understanding, however enormous, is generally limited.

Embarking on this adventure of untamed meals harvesting, arm your self no longer without a doubt with gadget however with attention. Invest time in knowledge nearby vegetation, attend workshops, are looking for the guidance of pro foragers, and at the equal time as uncertain, constantly refrain. It's better to miss out on a capability treat than to threat the wrath of nature's effective arsenal.

In the terms of a pro forager, "In the sector of wild edibles, you're both superb, or you're alive." This isn't always purported to discourage but to instill a revel in of reverence. For, inside the touchy balance of nature, wherein beauty and chance frequently wear the equal face, the actual paintings of foraging lies in recognizing the masquerades and celebrating the real.

The Forager's Ethic: Taking Without Depleting

Wandering amidst the verdant heart of the wooded area, in which nature unfurls in all her wild beauty, there lies a diffused settlement among man and nature. A silent p.C. Of understanding, of supply and take, in which the equilibrium remains undisturbed. This delicate balance forms the very ethos of a real forager: the art of harvesting bounties without rendering scars upon the landscapes that generously provide.

Imagine the historical woods as large libraries, wherein every plant, each mushroom, tells a story etched over eons. Now, think of a toddler moving into this library for the primary time, eyes good sized with marvel, hand attaining out to understand at each vivid tome at the shelf. This toddler's harmless hobby, if unchecked, can purpose illness, books scattered spherical, a few pages torn. It's an age-antique cautionary tale of methods unrestrained enthusiasm can supply accidental damage. Similarly, inside the barren region, the inexperienced forager, pushed with the resource of the amusing of discovery, may additionally inadvertently damage the very ecosystems they apprehend.

Nature, in her gigantic generosity, offers her treasures freely, but she has her rhythms, her cycles of regeneration. The sagacious forager is conscious that proper appreciation of nature isn't simply in the act of harvesting but in the facts of restraint. It's about knowledge that even as you stand amidst a patch of wild garlic or come upon a grove of ripe berries, it isn't always all intended for the taking.

How does one tread the thin line between appreciation and exploitation? It starts via the usage of embedding oneself deep within the rhythms of nature. When you pluck a wild herb, apprehend that taking the whole plant diminishes its risk of regrowth. By without a doubt snipping a detail, allowing the roots to remain anchored, you allow lifestyles hold its eternal dance.

Let's delve into the mushroom-studded geographical regions of the forest floor. Finding a patch of in shape for human intake mushrooms is much like stumbling upon a hidden treasure chest. Yet, the essence of ethical foraging isn't in filling one's basket to the brim however in making sure the mushrooms' spores are left in the lower again of to vow destiny boom. It's a delicate ballet of making sure continuity, of making sure that the morrow's forager well-knownshows as a good deal satisfaction in discovery as you probably did today.

And then there are the berries, nature's bejeweled candies. The pleasure of locating a bush weighted down with ripe berries is first-rate. But do not forget, we aren't the handiest ones who rely upon this succulence. Birds, bugs, and notable creatures partake on this ceremonial dinner. By making sure you're taking clearly enough, leaving within the returned of a honest percent, you're not certainly respecting nature, but you're furthermore weaving your self into the tough internet of lifestyles, as a player instead of an insignificant customer.

The ethos of sustainable foraging additionally translates into understanding the landscapes you traverse. Some regions, in particular those recovering from natural disturbances or those with uncommon species, require more sensitivity. It's about acknowledging that sure places are satisfactory left untouched, legitimate from a distance.

In the end, foraging, in its truest essence, isn't an insignificant act of collecting. It's a philosophy, an ever-evolving courting amongst guy and nature. It's about know-how that at the equal time as nature's pantry is tremendous, it isn't inexhaustible. Every plant taken, every berry plucked, leaves an imprint, however minute. And it is the collective weight of those imprints that shape the fitness of our wild regions.

The forests whisper reminiscences to individuals who listen. Tales of symbiosis, of existence in harmony. As foragers, our function isn't just to pay interest however to relate those testimonies via our movements, making sure that our footprints fade, but the tales live. So, the following time you assignment into the wild, basket in hand, permit it now not absolutely be an act of amassing, however a pledge. A pledge to take with out depleting, to rejoice without exploiting, and notably, to forge a bond with nature that's rooted in reverence, know-how, and eternal gratitude.

Respecting Local Ecosystems and Biodiversity

In the smooth consist of of barren area, in which every dewdrop sparkles with testimonies untold, and each rustling leaf narrates sagas of yore, lies the essence of lifestyles itself: biodiversity. As foragers, we're extra than mere web site visitors within the ones complex geographical regions; we're storytellers, custodians, and, most significantly, participants within the grand dance of lifestyles. But with this privilege comes the profound obligation of knowledge and respecting the very heartbeats of these ecosystems, ensuring our movements bolster in preference to wreck the touchy symphony of existence.

Step into any thriving wild place, and you will be serenaded with the resource of a charming cacophony of lifestyles. From the industrious ants tracing tricky paths on the wooded vicinity floor to the ethereal songbirds casting melodies from treetops, each organism performs a position, each which encompass a completely unique have a look at to the orchestra of existence. These man or woman notes, whilst woven together, create the wealthy tapestries of neighborhood ecosystems. And in the ones tapestries lie secrets older than time, secrets and strategies that educate us the actual which means of coexistence.

For the discerning forager, expertise close by ecosystems is similar to getting to know a new language. A language wherein each plant, each creature is a word, and expertise their interrelations, their symbioses, is to apprehend the nuances of grammar and syntax. It's information that the presence of a particular berry may symbolize the health of its pollinators, that a thriving patch of wild herbs might be the give up result of a specific soil pH nurtured thru nice decomposers. Every announcement, each nuance, is a clue, revealing layers of interwoven relationships.

However, this know-how is not actually intellectual; it's far deeply emotional and spiritual. It's feeling an notable surge of gratitude at the equal time as coming across a patch of wildflowers, information that those blooms are the quit end result of limitless interactions, from the sun's nourishing rays to the earthworms enriching the soil underneath. It's experiencing a profound experience of awe, understanding that every rectangular foot of undisturbed land is a thriving microcosm of existence, a universe unto itself.

Foraging, then, transforms from an insignificant act of collecting to a profound communicate with nature. A talk in which you do now not definitely take but supply back, ensuring your actions resonate with appreciate. It may additionally mean replanting seeds from surrender end result you devour, developing pathways for water to nourish parched areas, or sincerely refraining from stepping on a moss-protected log, expertise it's far a sanctuary for infinite tiny creatures.

Yet, respect for community ecosystems moreover implies acknowledging one's obstacles. Just as an artwork lover couldn't contact a masterpiece with out knowledge its fragility, a forager need to method ecosystems with the eye that they will be treading upon landscapes crafted over millennia. Disturbing a nesting ground or harvesting plant life that function number one meals property for nearby fauna can set into movement a cascade of disruptions, now and again irreversibly changing the balance of an environment.

Moreover, in a global in which the tendrils of city sprawl attain ever outward, and in which wild areas turn out to be more and more fragmented, respecting network biodiversity is no longer best a preference—it's an crucial. It's a clarion name to each forager, beckoning them to not sincerely be gatherers however guardians, advocates for the silent voices of the wild.

Chapter 6: Edible Landscapes And Habitats

Our planet is an ever-converting mosaic of landscapes, each proudly proudly owning its personal particular set of in shape to be eaten services. From the shadowy consist of of dense forests to the saline kiss of coastal marshes, or maybe within the bustling heart of our concrete metropolises, a plethora of culinary wonders look ahead to the discerning forager. Embracing the artwork of foraging is to adventure through the ones various habitats, unraveling their secrets, and celebrating the symphony of flavors they proffer. It's approximately the joys of discovery, the intimate connection with our surroundings, and the profound recognize for the bounty it gives.

Forest Foraging: Mushrooms, Nuts, and Leafy Greens

The cowl stretches overhead, a dense tapestry of leaves that filters the daytime into dappled patterns on the wooded region floor.

As you tread softly on a carpet of moss and decaying leaves, the whispering winds supply memories of historical instances when nature's pantry turn out to be open to all, and every wooded area became a bounteous larder. To forage in such environments is to step once more into those primeval times, looking for the equal wild edibles that sustained our ancestors.

Forests, with their complex ecosystems, host a plethora of culinary treasures. Among the maximum coveted are mushrooms. These fungi emerge from the loamy soil, often in the business enterprise of ancient wooden with which they percent a symbiotic dating. Morels, chanterelles, and boletes - each has its own precise flavor profile, from the deep earthiness of the truffle to the diffused nuttiness of the porcini. But caution is the forager's closest companion at the same time as looking mushrooms. For every suitable for eating delight, there can be a toxic doppelganger lurking close by. Proper identity is paramount, keeping apart the delectable from the lethal.

Yet, it isn't absolutely the fungi that beckon. As autumn breathes its golden colors into the wooded vicinity, nut-bearing wood drop their bounty. Chestnuts, walnuts, and hazelnuts, encased in shielding shells, wait to be amassed. Each nut holds a tale of the tree it got here from, shooting daytime, rain, and the very essence of the wooded region inside its kernel. Cracking open a sparkling walnut is just like unlocking a time pill of flavors, a mixture of the sweet and the earthy.

In the plush understory, leafy vegetables thrive, shielded with the aid of the taller timber. Dandelion leaves, sorrel, and nettle, regularly dismissed as mere weeds, are in truth nutritious and flavorful whilst prepared efficiently. Dandelion, with its slightly bitter notes, may be a smooth addition to salads or sautéed lightly for a thing dish. Sorrel, with its tangy citrus kick, presents zing to soups and stews. Nettle, although intimidating with its sting, turns into a mild, spinach-like green whilst cooked, wealthy in minerals and nutrients.

However, as the forest unveils its treasures, the forager need to additionally be attuned to its moods. Seasonality plays a important characteristic. While spring would in all likelihood herald the advent of soft veggies and early mushrooms, it is the cusp of autumn that in reality turns the forest right into a forager's paradise. Timing is critical, as is information of the particular habitats preferred via every healthy for human consumption. A damp, shaded grove would possibly possibly yield a treasure trove of chanterelles, at the same time as the sunlit clearings is probably wherein the nuts fall thickest.

Foraging within the wooded place isn't always honestly approximately accumulating meals. It's a communion, a talk with nature. Every rustle in the underbrush, each hen name, gives layers to the revel in. It hones the senses, making one observant of the diffused shifts, the minute facts - the moderate discoloration of a mushroom cap that indicators its edibility, or the precise shade of green that distinguishes one leafy green from its a whole lot less palatable cousin.

However, this immersion comes with responsibilities. Overharvesting can damage the delicate stability of the wooded area environment. Ethical foragers understand the want to take great what they may be able to use, leaving sufficient for nature to regenerate and for unique creatures to feed upon.

In the encompass of the forest, the traces some of the beyond and the prevailing blur. Each foraging day trip is a journey every outward into the verdant realm and inward into our primal selves, rekindling the historical bond we proportion with the land. The forest, in its silent information, teaches the forager to appearance, to pay attention, and most significantly, to apprehend. For in its shadowed depths lie now not just edibles, however memories, equipped to be located, savored, and shared.

Coastal and Wetland Treasures: Seaweeds and Water Plants

As waves crash on the shore, leaving within the lower again of a frothy residue, the coastal and wetland areas whisper a one among a type story of sustenance. Unlike the deep, muted tones of the woodland, the shoreline vibrates with a raw power, a rhythm set via the ebb and go with the flow of the tides. And it's internal this dynamic environment that some of the maximum exciting edibles reside, drawing lifestyles from the confluence of freshwater and saline.

Imagine taking walks on a wet, sandy beach truely after the tide has receded. Your feet could possibly brush towards the slippery fronds of seaweed, those marine algae that range from the translucent inexperienced sea lettuce to the robust, brown kelp. Seaweeds are nature's underwater gardens, swaying gracefully with the currents. Their culinary capability is large, boasting flavors as severa as their colors. Some, just like the nori, are subtly briny, reminiscent of the deep blue they arrive from. Others, on the side of the dulse, have an umami richness, making them perfect for soups and broths.

But it's far not without a doubt their flavor that is fascinating. Seaweeds are dietary powerhouses, brimming with minerals, nutrients, and antioxidants. They've been critical to the diets of coastal corporations around the globe, from the Irish who have loved carrageenan-weighted down moss for its gelling houses, to the Japanese who remodel wakame into sensitive salads.

Transitioning from the sandy seashores to the marshy nation-states of wetlands, we find out a selected set of treasures. These liminal zones, wherein land tentatively meets water, are teeming with life, every above and beneath the floor. Watercress, with its peppery bite, thrives in the cool, flowing streams, its roots submerged while its leaves reach for the solar. Then there can be the arrowhead, a plant aptly named for its triangular leaves, whose tubers are starchy morsels, often roasted or boiled with the useful resource of individuals who understand of their lifestyles.

Among the reeds and rushes, one may additionally find the water chestnut. Contrary to its call, it isn't a nut but an aquatic tuber that is crisp and subtly candy, a texture and flavor acquainted to the ones who've cherished Asian cuisines. But in all likelihood the maximum poetic of wetland edibles is the lotus. Rooted inside the dirt, its stems upward thrust via the water, culminating in blossoms that kiss the air. Every a part of the lotus is suitable for eating, from the younger seeds, tasting of sparkling coconut, to the mature roots that deliver a nutty earthiness.

Foraging in these moist terrains goals a heightened awareness. The transferring sands of the coast and the mucky grounds of wetlands can be treacherous. Tides are unpredictable, and what's to be had at one moment might be submerged the subsequent. Knowledge of tidal charts, an understanding of the lunar cycle's effect, and familiarity with the terrain are essential.

Furthermore, even as the ocean's generosity is big, it is not limitless. Overharvesting seaweeds can disrupt marine habitats, and taking too many tubers from a wetland can avert the boom of destiny flowers. Thus, the conscientious forager treads lightly, constantly making sure that they leave in the back of more than they take.

These coastal and wetland areas, with their specific edibles, are a testament to nature's adaptability. They're places of convergence, wherein factors collide and merge, ensuing in flavors as complicated as their origins. Each chew of seaweed or water plant is a flavor of the primordial soup, a nod to the waters from which all existence sprang.

To forage right right here is to have interaction with the primal forces of nature, to flavor the essence of the ocean and the freshness of flowing waters. It's a journey of discovery, wherein the street a number of the terrestrial and the aquatic blurs, and wherein each handful of harvest tells testimonies of historical seas, meandering rivers, and the timeless dance of the tides.

Urban Foraging: Surprising Edibles in Concrete Jungles

The metropolis's cacophony — blaring horns, far flung sirens, and the ever-gift hum of existence — may not strike one because the backdrop for a foraging day adventure. Yet, even in the metallic and glass canyons of our metropolis environments, nature persists, undeterred. Beneath the shadow of skyscrapers, within the not noted nooks of neighborhoods, the city panorama hides culinary treasures that challenge the convention of in which wild edibles live.

On a balmy afternoon, permit your senses manual you past the din. The fragrance of crushed herbs wafts from the cracks in a forgotten alleyway. Purslane, with its jade-like leaves, flourishes defiantly in the ones now not going areas. This succulent plant, regularly ignored as a weed, is a crisp, lemony satisfaction, best for glowing summer season salads or a zesty pesto. It's a testament to nature's resilience and flexibility, flourishing even inside the most hard terrains.

Turn a nook, and you could danger upon a mulberry tree. While they might be considered inconveniences via town safety due to their messy droppings, for the informed forager, the ones timber are a candy bounty. As summer season reaches its zenith, their branches sag with dark, juicy berries, harking back to raspberry but with a very unique tartness. Mulberries, whether or now not modified into jams or certainly popped into the mouth right off the department, are town gems prepared to be savored.

It's a peculiar appeal, this juxtaposition of the wild and the built. Urban apple wooden, probable remnants of antique orchards or the whimsical planting of a hopeful resident a few years in the past, offer aromatic blooms in spring and fruit in autumn. Hidden in the back of an vintage building or reputation solitary in a park, these timber undergo apples that may not have the polished appearance of grocery store types however supply flavors an prolonged manner more profound.

But it's miles no longer just the obvious, fruit-weighted down wood that beckon. The dandelion, regularly taken into consideration a gardener's nemesis, has smooth leaves that make for a barely sour however fresh salad green. Or consider the nettles, hiding in the underbrush of city parks, anticipating the careful hand to harvest and redesign them right into a rich, inexperienced soup or tea.

Strolling by a metropolis's network garden, one might also discover chives developing rebelliously out of doors their sure plots, their sensitive red blossoms equipped to be sprinkled over dishes for a gentle oniony zing.

The pursuit of city edibles is as plenty approximately connection as it's far about sustenance. It's a reminder of the land's facts in advance than it changed into paved and built upon. Every secure to consume weed or fruiting tree in the metropolis has a story, of seeds touring on the wind or being transported by means of the usage of using birds, of flowers that took root irrespective of the probabilities, and of nature's timeless urge to reclaim and thrive.

Of course, town foraging comes with its very own set of caveats. The issues of pollution and capability infection recommend that one desires to be selective approximately in which they forage. Proximity to busy roads or commercial employer areas, for example, may additionally render a spot less than first-class. A seasoned urban forager is conscious those intricacies, making sure that their harvest is each delicious and steady.

And then there may be the unwritten code — the understanding that genuinely because of the reality some thing grows in a public place does not recommend it is up for indiscriminate grabs. Respect for the vegetation and for fellow city dwellers is paramount. Harvest sustainably, taking quality what you could use, and typically making sure the plant can keep its boom cycle.

In the end, to forage in the town is to look it with new eyes. It's to recognize that even amidst the concrete and the chaos, there may be a pulse of existence, a green coronary coronary heart that beats regularly, offering its affords to those inclined to appearance. It's a adventure of discovery, wherein the road the various wild and the cultivated fades, and wherein each harvest is a party of nature's indomitable spirit.

Chapter 7: From Field To Plate

Foraging is not truly the act of gathering; it's far the bridge that spans the extremely good expanse amongst nature's bounty and our ingesting desk. While the thrill of discovery fills the coronary heart of the forager, the real magic unfolds while those wild treasures are translated into delectable bites, prepared to be savored. This transformation, from wild finds to culinary masterpieces, is an art work form in itself—a sensitive dance of care, creativity, and an knowledge of the uncooked essence of each issue.

Cleaning, Processing, and Storing Your Finds

The golden hours spent traversing forests, combing coastlines, or weaving thru town thickets yield treasures aplenty for the intrepid forager. Yet, as quickly because the a laugh of the quest ebbs, and you are left observing on the trove of wild edibles unfold earlier than you, the fact dawns: the ones raw, unprocessed provides of nature now beckon a modern day dance – the dance of training.

Take a 2d to revel in that first gaze upon your bounty. The serrated leaves, the lustrous berries, the craggy mushrooms – each with its tale, each looking ahead to a change. The metamorphosis from uncooked, untamed harvest to a pantry-prepared component is an artwork unto itself.

Begin with the water's caress. As you wash your gathered presents, agree with you are reintroducing them to their natural habitat – the babbling brooks, the cascading waterfalls. But this bath is not handiest a chic ritual. It gets rid of the dirt, the freeloaders, the residue of their wild beyond. As your hands skim the surfaces, come to be familiar with the textures, the anomalies, discerning the actual from the misleading. For in this tactile adventure, you could once in a while find out mimics, hangers-on, or a stray inedible nestled some of the right.

Next, we tread into the location of processing. Those sensitive vegetables you so carefully plucked? They may moreover find out contentment in a moderate blanching or a short sauté. Mushrooms, the ones capricious beings, ought to sing whilst grilled, their earthiness deepening with every caramelized word. And what of the radiant berries? Perhaps a mild simmer, wherein they burst proper into a cascade of tangy sweetness, is their future.

Remember, no longer all edibles call for transformation. Some whisper tales of centuries beyond and preference to live as they may be, an echo of the wilderness. But for people who need a touch of alchemy, ensure you honor their essence. Over-processing can mute their voices, making them a trifling shadow in their colourful selves. Like a maestro, orchestrate this symphony with care, ensuring each have a look at — be it a steam, a roast, or a overwhelm — elevates the thing.

Then comes the paintings of upkeep. How do you seize a fleeting season in a jar? How do you make certain that the vibrancy of spring or the richness of autumn lingers thru the frost and bloom? Storing your finds is greater than mere refrigeration or shelving. It's about growing a time tablet.

Certain vegetables could probable fancy a groovy, dry corner, their crispness intact for days on give up. Berries, in the meantime, would probable looking for the cold consist of of a freezer, their vibrancy locked in a frosty vault. Mushrooms, those fickle entities, need to preference a quick sunbath before nestling in a groovy, dark alcove. The art is in understanding, in listening, in records the silent dreams of your harvest.

And then, as days come to be nights and seasons ebb and go along with the flow, at the same time as you obtain out for that jar of solar-dried tomatoes or the frozen pouch of springtime berries, you will be transported. Back to that 2nd below the quilt, thru the coastline, or amidst the city sprawl. It's now not just about sustenance; it's miles approximately reliving a memory, a 2d, a fleeting emotion.

In the grand tapestry of foraging, the act of cleansing, processing, and storing isn't a postscript. It's a economic catastrophe brimming with its very personal adventures, its own memories, its private magic. For within the ones acts, you aren't simply preserving meals; you are keeping testimonies, moments, and the very essence of the wild.

Culinary Delights: Simple Recipes to Start With

Nature's bounty, once foraged and prepped, is a canvas of limitless culinary possibilities. The wild flavors, the untamed aromas, and the unrestrained textures, all integrate to bop a primal jig at the palate. Yet, the splendor of those wild components is that they do not commonly call for complexity in cooking. Sometimes, the most effective recipes are folks that permit their proper essence shine the brightest.

Imagine a overdue afternoon. The solar, a golden globe, caresses the horizon. You discover your self in a kitchen bathed in its warm embody. On your countertop lie a handful of untamed berries, some freshly picked dandelion veggies, and a gaggle of chanterelle mushrooms. It's a gathering of nature's nice, and you're approximately to weave culinary testimonies with them.

Let's begin with those wild berries, bursting with juiciness and a symphony of candy and tart. Perhaps a compote? A mild simmer with a sprinkle of untamed honey, a dash of easy spring water, and a whisper of lemon zest. As it bubbles, the berries rupture, liberating their colourful colorings and extreme flavors, melding right right into a luscious, velvety compote.

Drizzle it over home made pancakes, or allow it enrobe a scoop of vanilla ice cream. The flavor? Pure, wild, and unadulterated magic.

Now, those dandelion vegetables. Often unnoticed, however they're a green with grit, with an earthy robustness that holds its personal. How about a country pesto? The veggies, coarsely chopped, combined with wild garlic, pine nuts, and bloodless-pressed olive oil. Blend them until they merge right into a verdant, aromatic paste. Toss it thru some freshly boiled pasta, or smear it on a crusty slice of sourdough. Each chunk is a homage to the meadows and forests from whence they came.

And then the chanterelles, golden and aromatic. These fungi are nature's umami bombs. Think of a easy sauté. Melt a few wild butter in a skillet, and because it froths, toss within the mushrooms. The sizzle, the aroma—it is sensory satisfaction. A sprinkle of sea salt, a grinding of pepper, and a few sprigs of wild thyme. Let them prepare dinner until they're caramelized, their flavors intensified, their textures crisp yet smooth. Serve them atop a slice of toasted brioche or as a factor to a grilled steak. It's the essence of the wooded location on a plate.

Yet, the adventure does now not prevent proper right here. The global of wild edibles is huge, and the recipes are however a drop on this ocean. The key is to include simplicity. To permit the components sing their songs, narrate their recollections. Overcomplicating can from time to time overshadow their innate beauty.

It's also about intuition. While recipes provide a framework, it is the coronary coronary heart, the soul, and the senses that guide the real culinary dance. It's approximately feeling the element, statistics its whispers, and knowing its goals. Sometimes it's far a moderate roast, on occasion a brisk blanch, and at times, only a raw, untouched presentation.

The splendor of those smooth recipes is the reminiscences they tell—the tale of the land, the seasons, the factors. Each bite is a adventure, every taste a memory. Whether it is the splendor of the berries reminding you of a sunlit clearing, the earthiness of the dandelions taking you returned to a dew-kissed meadow, or the robustness of the chanterelles echoing the woodland's depths.

Embracing wild factors in a single's culinary repertoire is more than just an act of cooking. It's an act of reverence. It's approximately honoring nature, cherishing its presents, and celebrating its abundance in the maximum intimate manner possible—through savoring it. So, the following time you discover your self with a trove of foraged delights, take into account, simplicity is top. And allow the wild manual your culinary muse.

Pairing Wild Ingredients: Flavors of the Wild

In the symphony of gastronomy, pairing components is much like merging the melodies of character devices. The wild worldwide of foraged food presents a very precise timbre, a legitimate that resonates with nature's rhythms. These flavors, strong and untamed, require a sensitive ear, or as an alternative, a discerning palate, to mixture in harmonious live normal overall performance. When completed proper, the stop end result is an opus that captivates the senses and transports one to forgotten landscapes.

Consider the earthy, nearly mystical nice of wild desserts. On their very very personal, they may be a notable testament to the forest's depth, but mixed with the proper accomplice, their majesty amplifies. Imagine marrying them with the velvety richness of a gradual-cooked egg yolk. The truffle's aromatic oils meld with the yolk's creamy texture, developing a fusion it truely is every grounding and elegant.

Yet, the art work of pairing isn't about pressure or contrivance; it's about listening. It's about knowledge the whispers of the wild blackberry because it yearns for the tang of wild goat cheese. It's feeling the pull of the salty samphire within the path of the buttery decadence of scallops. Each wild component, whether or not plant, fungi, or fauna, has a story, and our challenge is to allow those memories to entwine, to dance, to sing collectively.

Take, for instance, the dandelion greens with their peppery bite. To a few, they'll echo the sunlit meadows, the brisk wind, the open sky. But within the realm of pairing, they reap out for a counterbalance. A drizzle of honey possibly, or the smoothness of avocado. Together, they invent a juxtaposition, a pleasant assessment wherein one elevates the other.

The key to unlocking the ones pairings lies not in grandeur or flamboyance but in subtlety. It's not about crowding the plate or the palate. It's approximately providing area, allowing every flavor to respire, to particular itself. It's approximately recognizing that once in a while, a wild nettle ought to no longer want an complex sauce. Maybe all it seeks is the encompass of a nutty brown butter or the gentleness of a sparkling ricotta.

Yet, this dance isn't always quite an entire lot flavor. It's a multisensory enjoy. The aroma of untamed mint could in all likelihood evoke memories of riversides and picnics. Paired with the zing of untamed lemon zest, and all at once you're not definitely tasting but moreover reminiscing, feeling, and traveling via time and area.

Of course, venturing into the wild international of pairing is not with out its demanding situations. There might be missteps. The sturdy muskiness of advantageous wild mushrooms should possibly battle with an further assertive wild herb. But even in the ones moments, there can be a lesson, a whisper from nature announcing, "Not this manner, try some different." And as with any top notch symphonies, exercise and endurance cause perfection.

One may moreover surprise, why undergo this elaborate dance? Why now not permit the wild components shine solo? The answer is each smooth and profound. Pairing is a party. It's a nod to the interconnectedness of nature. It showcases that even within the wild, there's a rhythm, a balance, a enjoy of belonging. Every wild component, in its essence, is trying to find its companion, its counterpart to create a fuller, richer narrative.

As the sun dips below the horizon, casting the arena in colours of gold and crimson, undergo in mind sitting proper right down to a meal. A meal in which every bite tells a tale, wherein every flavor pairing is a duet of wild voices making a song in concord. It's not simply sustenance; it's an revel in. An ode to the wild, to nature's vastness, to its intricacies, and to its profound simplicity.

In the cease, the act of pairing wild factors is a adventure. A journey of discovery, of ardour, and of reverence. It's about spotting the profound splendor within the wild and showcasing it in a way that touches now not truely the tongue however additionally the soul. It's about echoing the eternal dance of nature on our plates. And in doing so, it reminds us of our very own wild essence, our very personal vicinity in this tremendous, lovable tapestry of existence.

As we adventure from the sphere to our plate, we are reminded of the beauty in nature's services and our role in honoring them. It's now not quite lots growing a meal; it is approximately celebrating the untamed, the unbridled, and the sudden. As every wild component famous its area in our culinary creations, we no longer simplest satisfy our hunger however furthermore pay homage to the land and its boundless generosity.

Chapter 8: Medicinal And Practical Uses

The splendor of untamed flora extends far past their potential to satiate starvation. They weave a rich narrative of recuperation, properly being, and ingenious expression that has resonated through time. As we meander through existence's many paths, nature continuously gives remedies to appease our ailments, factors for our fitness physical activities, and substances to kindle our innovative flames. By diving into this profound relationship among people and the wild, we unearth practices which can be each age-antique and refreshingly novel, revealing the multifaceted wonders that the herbal global bestows upon us.

Herbal Remedies from Common Wild Plants

The earth, in all her unfathomable information, has provided no longer best nourishment for our our bodies but moreover effective remedies for our myriad illnesses. Meandering through a wild meadow or wandering through the usage of a woodland's component, one is frequently amidst nature's personal pharmacy. Far from the pristine shelves of modern-day apothecaries, the ones remedies develop freely underneath the solar and stars, searching beforehand to discerning palms to reveal their restoration secrets and strategies.

Dandelion, frequently disregarded as a mere weed, graces many landscapes with its sunny disposition. Yet, its roots and leaves are treasure troves of nicely-being, known for loads of years as liver tonics. A mild infusion can offer a fulfilling, sour tonic that stimulates digestion and detoxifies. This humble plant is also a diuretic, helping in eliminating greater water from the body without depleting it of essential minerals.

Plantain, no longer to be pressured with the tropical fruit, is a not unusual sight in gardens and paths, its rosettes of leaves clearly left out. Yet, while implemented as a poultice, it really works wonders on insect bites, drawing out venom and lowering contamination.

Moreover, a tea made from its leaves can soothe an irritable bowel, relieving soreness with its mild anti inflammatory homes.

Red clover, with its trifoliate leaves and red plants, is not best a beauty to behold. It's a plant deeply woven into ladies's health. Used as a tea, it can assist alleviate signs and symptoms of menopause and is wealthy in isoflavones, compounds seemed to have antioxidant houses.

Yarrow, the diffused white vegetation status tall in many grasslands, has been a staunch great buddy of healers for millennia. Its Latin call, Achillea millefolium, pointers at its legendary records; it's far stated that Achilles, the Greek hero, used it to cope with infantrymen' wounds. This recognition isn't unfounded; yarrow can staunch bleeding and disinfect wounds. Moreover, taken as a warmness infusion, it is able to help reduce fevers via selling sweating.

Mullein, with its towering yellow plant life and gentle leaves, frequently graces disturbed soils, roadsides, and fields. Its leaves, whilst steeped right right right into a tea, can act as a demulcent, soothing angry membranes, particularly inside the respiration system. It's a slight remedy for coughs, especially dry ones that rattle the chest.

Yet, with all this potential, a word of warning should be interjected. The line among medicine and poison can occasionally be slender. Comfrey, for instance, can be each a treatment and a toxin. While its outdoor use in poultices can beneficial resource in bone recuperation, its internal intake can damage the liver, way to wonderful alkaloids. Thus, information and respecting the efficiency of untamed herbs is paramount.

Additionally, sourcing performs a pivotal role. A plant's medicinal nice can be encouraged via in which it grows. Those through the roadside may also additionally have absorbed pollutants, while ones from pristine environments would possibly likely offer a higher recovery charge.

While current medication offers precious answers, reconnecting with those historic, herbal treatments may be an enriching revel in. It's about reviving an ancestral bond, a whispered legacy surpassed thru infinite generations. Every sip of dandelion tea or utility of a plantain poultice is a nod to our forebears who, below canopies of green or by flickering firelight, first unraveled the recovery mysteries of the verdant international around them.

Crafting Nature-Based Skincare and Wellness Products

The encompass of the forest, the mild rustle of meadow grasses, and the far flung hum of a river deliver to mind an historical alchemy, an artwork deeply woven with the tapestry of the earth. This artwork lets in us to attract from nature's bounty and craft skin care and health products that rejuvenate the frame and echo with the very essence of the wild.

Imagine beginning your day, not with industrially-produced lotions, however with a balm made of untamed rose petals, handpicked at sunrise while they will be despite the fact that cradled via dew. Such a balm does not just moisturize; it includes the perfume and mystique of untouched meadows. To create this, smooth rose petals can be steeped in a base oil (like almond or jojoba) in a pitcher jar, left in a sunlit spot for some weeks, permitting the petals to infuse the oil with their essence. Strained and combined with herbal beeswax, this consequences in a balm that is each nourishing and aromatic.

Consider calendula, a radiant bloom paying homage to the solar. Its petals, at the same time as infused in oils, were recognised for their calming homes, particularly for irritated pores and pores and skin. One may craft a salve by using grinding dried calendula petals proper into a powder, and then lightly simmering it in a carrier oil. Once strained, this oil, mixed with shea or cocoa butter, creates a salve that may be a balm for distressed pores and pores and skin.

But nature's treasures amplify beyond virtually pores and pores and skin care. Think of bathtub salts infused with the essence of pine needles or cedarwood chips. This isn't genuinely approximately a calming bathtub; it is a dive right right right into a woodland flow into. One ought to collect fallen pine needles, dry them, and then combination them with Epsom salts and some drops of cedarwood important oil. The give up result? A bathtub that looks as if a wooded region retreat.

And at the times even as strain knots your muscle companies, a rubdown oil infused with wild chamomile or St. John's wort is probably the panacea. These wild herbs may be cold-infused in a carrier oil over numerous weeks. Once organized, the strained oil may be a base for healing rubdown blends, melting away tension and weariness.

Crafting those products, despite the fact that, is not merely about following recipes. It's statistics the rhythm of nature, the person of each plant, and the manner their essences can be fine drawn out. Some herbs are exceptional used glowing, whilst others screen their efficiency while dried. Some might probably want the warmth of the solar to release their magic into infusions, while others, like crucial oils, also can require steam distillation.

But as one walks this direction of wildcrafting, a word of caution: nature is robust. Just because of the reality a few element is natural does now not imply it's miles benign. Always do a patch take a look at earlier than attempting out new concoctions. And, don't forget the requirements of sustainable foraging; take only what you need, ensuring you do not disturb the delicate stability of nature.

In the surrender, as you hold a jar of wildcrafted cream or a bottle of natural oil, apprehend that you're protective extra than only a product. It's a connection, a tale, a bridge among you and the big, generous coronary heart of the wild.

Wild Plants in Traditional and Contemporary Crafts

As the solar dapples the wooded area ground and the songbirds serenade the dawning day, one cannot help however enjoy a profound connection to the land below their ft. And in that hallowed connection lies an age-antique way of lifestyles: crafting with wild flowers. While many see the barren region as a pantry of edibles, it's also an artist's palette, teeming with substances and idea for every traditional and contemporary crafts.

Think of the trendy cattail, which stands tall along the water's region. For generations, these were greater than simply visual delights. Native communities wove them into mats and baskets, growing devices that had been each useful and aesthetic. Each woven strand recommended a tale, each loop a testomony to nature's bounty and the crafter's potential.

And what of the smooth willow? Its slim, pliable branches, or withies, have danced in the arms of many a craftsperson. Traditional willow weaving techniques starting baskets, sculptures, or even living systems that blend seamlessly into herbal landscapes. As the willow tendrils spiral and intertwine, they'll be not just forming shapes; they may be echoing the rhythms of the wild, mirroring the serpentine trails of rivers or the slight spirals of unfurling ferns.

Yet, it isn't really in age-antique traditions that wild plant life have placed their voice. Contemporary artists are embracing the wild, crafting avant-garde installations and artistic endeavors. Picture a present day-day studio in which moss, generally seen carpeting the forest ground, turns into a residing tapestry. This lush, verdant wall is not really ornamental. It's alive, respiratory, a poignant statement on nature's quiet resilience.

Similarly, many artisans are turning to dyes sourced from the wild, bringing the colours of meadows and woods to fabric. Whether it's the deep blues derived from indigo or the fiery oranges extracted from madder roots, the ones colours do not just tint; they sing, they communicate, they whisper testimonies of moonlit nights and sunlit glades.

One of the most evocative examples is probably sculptures made from driftwood. Each piece, worn and original by using way of nature's forces, carries the echo of crashing waves and the reminiscence of faraway seashores. Artists see beyond their gnarled, weathered exteriors, locating bureaucracy and figures geared up to be unveiled. A piece of driftwood, within the palms of a visionary, would possibly probable emerge as an imposing eagle in flight or a touchy nymph, every grain and curve telling its unique saga.

However, on the equal time because the desolate tract gives a treasure trove, it is important to technique it with reverence. Overharvesting or unfavorable habitats for craft substances is as grievous as doing so for meals. Crafters ought to attune themselves to the land's rhythm, records whilst to take and even as to provide once more.

Moreover, crafting with wild materials isn't pretty a high-quality deal the give up product. It's a adventure, a meditation. Each plant has its temperament, its quirks. To craft with them is to interact in a speak, to pay attention and to respond, to understand and to conform. It's about staying energy, approximately permitting the materials to guide the fingers, in vicinity of enforcing one's will upon them.

Chapter 9: Building Community Around Foraging

Foraging, an historic exercising of accumulating food from the wild, is more than most effective a manner to satiate hunger—it's far a bridge connecting humans to nature and to each extraordinary. Over time, as metropolis sprawls rose and digital displays ruled, this elemental connection risked being overshadowed. Yet, via communal obligations, workshops, and the timeless artwork of storytelling, the foraging community has decided techniques to boost and amplify its bond. It's inside the ones shared critiques and exchanged narratives that the real essence of foraging as a communal act shines brightest.

Organizing and Attending Foraging Workshops

The sun dapples the floor as whispers of pleasure rustle thru the bushes. People, with baskets in hand, come together to remedy nature's secrets and techniques and techniques within the heart of the wooded area. Welcome to a foraging workshop, wherein age-vintage traditions intertwine with newfound friendships, making the act of sourcing meals and remedy from the wild a collective adventure.

Foraging workshops are extra than certainly occasions; they'll be sanctuaries of reading and bonding. A region in which metropolis dwellers, frequently disconnected from their green roots, find a bridge to nature. Where the nuances of the wild worldwide are unraveled now not genuinely thru books or memories but via the tactile experience of touching, smelling, and tasting.

When one gadgets out to put together such a amassing, the most motive is to create an environment of inclusivity and recognize — for each the individuals and the land. Picking the proper place is paramount. While deep forests or secluded meadows can also moreover sound idyllic, remember accessibility for attendees of all ages and mobility levels. Spaces with a mixture of terrains, from open fields to wooded regions, provide a entire experience with out making it exhausting.

Next, rope in a close-by professional, someone with a profound knowledge of the location's vegetation, who can get rid of darkness from the world of foraging past mere identity. This person want to personal the air of mystery to captivate an target market and a coronary coronary heart that beats for the well-being of the barren region. Their function is pivotal in guiding attendees, making sure they tread lightly, only taking what's sustainable, and leaving no trace in the lower back of.

Workshops aren't quite a first rate deal gathering; they're approximately connecting. To foster a feel of community, incorporate sports activities that inspire interaction. Perhaps a consultation wherein attendees percentage their private foraging reminiscences or a project to prepare a wildcrafted meal with the day's haul. These moments, interspersed with reading, make for memories that linger lengthy after the day's sun has set.

For those on the opportunity aspect – the attendees – foraging workshops provide a treasure trove of stories. Beyond the enchantment of learning new abilities, attending a workshop may be a transformative journey. The current-day global, with its relentless pace, frequently leaves us longing for real human connections. What higher vicinity to locate them than beneath the expansive sky, surrounded by way of the usage of the use of the symphony of nature, working together to unearth nature's bounties?

But it's miles crucial to attend with an open coronary coronary heart and thoughts. Embrace the possibility not truly to build up understanding however to percent your private, to not simply take from the land however to provide decrease decrease again in gratitude. Be prepared for surprises - possibly the invention of an tremendous herb or the serendipitous forging of a lifelong friendship.

As dusk wraps the day, the attendees collect round a campfire, their faces illuminated with the aid of manner of manner of its heat glow, baskets brimming with the day's harvest. Stories are exchanged, laughter flows, and melodies of human beings songs fill the air. It is proper right right here, amidst this melding of beyond and gift, that the spirit of network round foraging is simply kindled.

Foraging workshops act as ties that bind humans to every different and to nature inside the massive fabric of existence. They feature proof that becoming a member of forces as a community may additionally additionally growth the fun of discovery and the depth of statistics in our search for nourishment, whether or not or no longer it's physical or emotional. So maintain in thoughts that it is no longer exceptional approximately what you discover in the wild; it's also approximately the wild relationships you create with the land and its populace, whether or not you're web hosting a workshop or attending one.

Foraging Groups, Festivals, and Community Projects

Beneath the vastness of the azure sky, recall a meadow pulsating with colourful colours, tune drifting within the wind, and a kaleidoscope of human beings accrued with a shared cause: celebrating the age-old art work of foraging. These aren't mere gatherings; they may be the soulful heartbeats of groups coming alive in their energy of will to nature and historical past.

Foraging companies are an alchemy of historical information and present day-day camaraderie. Typically initiated with the beneficial useful resource of a handful of lovers, these businesses evolve into dynamic collectives, wherein amateur foragers have a look at from pro professionals, and traditions skip from one generation to the following. Each excursion, whether or not or no longer to a verdant wooded place or a golden coastline, is an tour into the coronary coronary heart of nature and an exploration of human connection. Every root unearthed, every berry picked, turns into a story, a photograph of the corporation's shared reviews.

Yet, past the ones normal tours, the spirit of network culminates in foraging fairs. These are grand, jubilant events, celebrating nature's bounty. Set in picturesque locales, the ones festivals are a panorama of sports activities, from guided foraging walks to culinary contests, wherein wild factors take center degree. Artists craft tunes inspired with the useful resource of way of rustling leaves and chirping birds, while kids take part in workshops, their hands stained with berry juice, eyes sparkling with marvel.

Amidst the merriment, a competition often serves a deeper reason: to enlighten. Exhibits and demonstrations on sustainable foraging practices make certain that attendees understand the delicate stability of nature. Renowned foragers percent stories of their maximum enigmatic discoveries, at the identical time as ecologists elucidate the symbiotic dating amongst humans and the environment. In those moments, the competition transforms from a mere event to a profound academic experience.

Parallelly, network projects rooted in foraging spring forth, weaving the thread of sustainability into the cloth of society. Take, for instance, metropolis foraging maps, created with the beneficial aid of and for the community. These charts manual metropolis dwellers within the path of wallet of untamed edibles nestled amidst the concrete sprawl. Such initiatives no longer only sell community foraging however moreover highlight the importance of town green areas and their conservation.

Another heartening agency is the reputation quo of network gardens. Once barren plots of land metamorphose into thriving ecosystems, bearing surrender result, herbs, and vegetables. They stand as oases on the town settings, wherein humans come together to till, plant, and harvest. Each sprout that pushes via the soil, every tomato that ripens under the sun, is a testomony to the network's collective attempt. These gardens come to be epicenters for foraging workshops, wherein locals discover ways to come to be aware about suitable for consuming flowers and harness their blessings.

While fairs have fun and corporations discover, network duties empower. They allow foragers to offer lower back, to ensure that the wild areas, that have generously shared their treasures, flourish for generations to return. And in this project, every man or woman plays a pivotal role, whether or not or not or now not with the aid of the usage of planting a sapling, charting a modern foraging path, or really sharing the ethos of sustainable foraging with their friends.

In essence, foraging in its purest form is greater than a solitary quest for wild edibles; it's a symphony of voices, palms, and hearts coming collectively. The corporations provide a sanctuary for shared gaining knowledge of, the gala's a platform for joyous birthday celebration, and community duties a course inside the direction of sustainable living.

In a worldwide in which virtual screens often overshadow verdant greens and in which immediately gratification eclipses the delight of anticipation, those collectives remind us of the primal, profound connection we percent with the earth. They beckon us to step out, to revel in the soil below our feet, the wind in our hair, and to be part of some thing massive than ourselves.

As the sun devices on a pageant day, casting a golden glow over a meadow packed with laughter, track, and dance, one realizes that foraging isn't pretty tons what we take from the land. It's approximately what we give all over again – in gratitude, in strive, and in spirit. It's about building a community that respects, cherishes, and celebrates the wild wonders round us.

Sharing Stories: The Forager's Tale

In the heart of every forager beats a story, a personal chronicle imbued with marvel, discovery, and reverence for the land. The forager's story is a woven tapestry of adventures and epiphanies, of quiet moments beside babbling brooks and exhilarating ones while a sought-after mushroom is located. Each story, precise in its cadence and emotion, bureaucracy an indelible a part of the grand anthology of humankind's courting with nature.

Envision Clara, a septuagenarian, her eyes the deep hue of wealthy soil, her arms gnarled just like the roots she so frequently seeks. With each wrinkle on her face, there may be a story to inform. Perhaps she remembers a day from her kids whilst, guided by the usage of her grandmother, she positioned her first patch of wild strawberries, their scarlet brightness contrasting in opposition to the green, their flavor a sweet burst of sunshine. That modified into the day, Clara can also moreover say, she virtually understood the because of this of nature's generosity.

Then there may be Jamal, a town dweller have turn out to be passionate forager, who may also regale listeners together along together with his story of stumbling upon a grove of elderflowers within the coronary heart of an town park. The discovery, for him, grow to be a revelation—nature's bounty might be found even inside the steel and urban jungles of modernity. It became a name to sluggish down, to test, and to reconnect.

In each tale, there's greater than only a recounting of activities; there's an emotion, a sentiment, a lesson. Whether it's the joy of discovery, the a laugh of the chase, or the profound connection felt to ancestors who too walked the land and foraged for sustenance, the ones reminiscences provide a window into the very soul of the forager.

But why is the act of sharing those stories so vital?

For one, testimonies bind agencies. When a seasoned forager shares stories in their exploits, they may be not sincerely recounting sports; they may be passing down knowledge. The locations wherein wild garlic flourishes, the manner the earth smells without a doubt earlier than morels start to seem, the mild difference in hue amongst an fit to be eaten berry and its toxic lookalike—those styles of nuggets of statistics get transferred via tales. They feature every training and caution, making sure that traditions and strategies, honed over generations, maintain to thrive.

Chapter 10: Global Foraging

In the grand theater of Earth's landscapes, there may be a non-prevent play of flavors, textures, and aromas waiting to be unearthed. Every continent, with its one-of-a-kind ecology and way of life, gives wild edibles that mirror its soul, information, and culinary historical past. From the sun-baked terrains of the Americas to the historical soils of Europe and Asia, then onward to the vastness of Africa and the island paradises of Oceania, foraging is a adventure that transcends borders. It's a discovery of tastes so profound and narratives so tricky that they inspire a renewed appreciation for nature's bounty and human ingenuity.

The Americas: From Prickly Pears to Maple Syrup

The Americas, spanning from the Arctic tundra to the windswept pampas, harbor a tantalizing collection of untamed edibles which have been each a sustenance and a story for its population. Envision for a second, the majestic expanse of the land, in which every corner has an untold secret, a flavor that is ready to be determined, tasted, and celebrated.

Venture southward into the arid deserts of North America, and you could come upon the cactus, a stoic sentinel of the sands. Among them, the prickly pear stands proud not handiest for its thorny demeanor however for the candy, water-rich fruit it hides. To the indigenous peoples of the region, this fruit, known as "tuna", isn't always simply meals however a lifeline, a reservoir of hydration, fighting the parched environment. But it isn't absolutely survival; it's about flavor. A chunk right into a ripe prickly pear is a burst of sweetness, a touch of melon, and a hint of the wild.

As we traverse the continent, transferring in the path of the japanese woodlands, a particular tale unfolds. Here, under the snowy cowl of wintry climate, lies a treasure each sweet and liquid - maple syrup. The art work of extracting sap and converting it into this amber nectar dates lower lower returned centuries, a craft honed with the aid of the indigenous tribes and later followed by means of way of manner of European settlers. The sugar maple tree, in its silent statistics, offers up its lifeblood, which whilst decreased, brings forth a sweetness that is rich and fantastic. But beyond its culinary appeal, the machine of "sugaring" will become a communal affair, an event in which households collect, percentage recollections, and characteristic a super time the primary signal of the imminent spring.

Between the ones iconic flavors, lies a plethora of tastes that outline the full-size culinary landscape of the Americas. In the immoderate-altitude plateaus of the Andes, wild tubers like oca and ulluco are unearthed, their earthy flavors a staple for the mountain-dwelling agencies. The lush rainforests of the Amazon basin cowl limitless end result, some familiar like the guava, others lesser-mentioned but similarly tantalizing much like the camu camu, brimming with food plan C and a zesty tang.

Even in regions which can appear inhospitable, the land generously offers its bounty. The Atacama, one of the driest deserts on Earth, surprises with its wild herbs and roots, tailored to the harshness but bearing flavors which might be deeply invigorating. Or the Arctic tundras, in which underneath the cold facade, berries like cloudberry and lingonberry ripen, their tartness a adorable evaluation to the stark surroundings.

However, foraging in the Americas is not quite tons the man or woman flavors but the symphony they invent collectively. It's the melding of the candy, the bitter, the earthy, and the tangy. It's the guacamole made from wild avocados, the tangy refreshment of a wild lime drink, or the comforting warmth of a soup made with handpicked morel mushrooms.

What stands proud, as we adventure from one nook of the continent to the opportunity, is the deep-rooted connection the numerous land and its human beings. Every foraged item consists of with it the legacy of the tribes, the households, and the man or woman foragers who have walked the land earlier than. It's a testament to the honour for nature, the facts of its rhythms, and the pleasure of discovering its secrets and strategies.

In the stop, the wild edibles of the Americas are not sincerely substances; they're narratives, testimonies of survival, party, and community. They beckon us to appearance nearer, to taste, to get pride from, and to enroll in inside the age-antique dance of foraging, a dance that tells the tale of a land, large, numerous, and vivaciously flavorful.

Europe and Asia: Truffles, Bamboo Shoots, and More

Journeying from the grand boulevards of Paris to the misty bamboo forests of Sichuan, the historical lands of Europe and Asia unfurl a enthralling panorama of flavors. Each location, every valley, or maybe every hidden wooded area glade includes within it a storied factor, every extra charming than the ultimate.

Let's traverse this enormous expanse, weaving via cultures and landscapes, to find out the wild edibles which have original civilizations and tantalized palates for millennia.

In the picturesque location of Périgord, France, a gastronomic treasure silently matures below the dappled color of okaytrees: the truffle. Often termed the "diamond of the kitchen", this fungi's charm isn't always honestly its scarcity but the intensity of flavor it brings. A mere shaving of this delicacy can growth a dish from mundane to extremely good. Truffle looking, with expert dogs and a keen instinct, will become nearly a poetic organisation, a sensitive dance among guy, nature, and the elusive treasure beneath.

Transitioning from the European woodlands to the rolling meadows of Anatolia, wild herbs like sage, thyme, and oregano scent the air, every carrying memories of historical empires and nomadic tribes. They've been staples in kitchens, each humble and royal, consisting of layers of flavor to dishes and infusions.

Moving eastwards, as we climb the rugged terrains of the Himalayas, we are greeted by the usage of the stinging nettle. While its sting may probably deter the uninitiated, local communities have prolonged identified its secrets and techniques. When cooked, the edge dissipates, revealing a leafy inexperienced it truly is not simplest scrumptious but complete of nutrients. It's a testament to nature's paradox: on occasion, what appears forbidding in the starting appearance conceals a coronary heart of gold.

But it is in the coronary heart of Asia, amidst the dense bamboo groves, that a seasonal surprise emerges: bamboo shoots. In places like Japan, China, and India, the arrival of gentle bamboo shoots alerts the onset of spring. Harvesting them is an paintings, wherein timing is the entirety. Too early, and they may be too clean; too overdue, and they grow to be too woody. But whilst picked virtually proper, they lend a crispness to dishes, a subtle taste it really is reminiscent of the woodland after a spring rain.

Alongside the ones edibles, Europe and Asia have moreover supplied drinks that captivate. Consider the samovars of Russia, wherein wild berries and herbs are brewed into fragrant teas, or the traditional Japanese matcha, in which wild inexperienced tea leaves are floor proper right into a awesome powder, taking pictures the essence of the land in a cup.

One cannot delve into Asia's foraging wonders without pausing on the banks of the Mekong, in which the water spinach flourishes. Floating markets brim with bundles of this aquatic vegetable, its clean leaves and crisp stems a cherished aspect of many a Southeast Asian dish.

Yet, for all of the splendid flavors, there's a thread of universality. Be it the goulashes enriched with wild paprika in Hungary or the Korean kimchi made piquant with wild garlic, there can be an underlying apprehend for the land. A profound expertise that nature, in her limitless information, offers materials that aren't truly sustenance however memories, ready to be shared.

Chapter 11: Identifying Edible Wild Plants

Foraging for edible wild plants is a nuanced art that involves not only the thrill of exploration but also a deep understanding of plant morphology. In this section, we delve into the intricacies of identifying various parts of plants—leaves, flowers, seeds, roots, fruits, and stems—providing detailed information on safe identification and offering tantalizing recipes to transform your foraged finds into delectable dishes.

Leaves: Recognizing Safe and Nutrient-Rich Foliage

Leaves constitute a significant portion of edible wild plants, showcasing an remarkable diversity of shapes, sizes, and textures Before plucking any leafy greens for consumption, it's crucial to ensure accurate identification to avoid potential toxic look-alikes.

Identification Tips:

Leaf Shape: Examine the overall shape of the leaf, including its margins (edges) and any distinctive lobes or notches.

Vein Patterns: Pay attention to the arrangement of veins on the leaf surface, as it can vary widely between species.

Leaf Texture: Touch and feel the leaves. Some may be smooth; while others could be hairy or have a unique texture.

Color and Size: Note the color of both the upper and lower leaf surfaces additionally, observe the size and arrangement of leaves on the plant.

Recipes:

1. Wild Greens Salad:

Ingredients: Assorted edible wild leaves, olive oil, balsamic vinegar, salt, pepper.

Instructions: Wash and mix various wild leaves Drizzle with olive oil and balsamic vinegar, season with salt and pepper.

2. Nettle Soup:

Ingredients: Nettle leaves, potatoes, onions, garlic, vegetable broth, cream, salt, pepper.

Instructions: Saute onions and garlic, add chopped potatoes and nettles. Pour in vegetable broth, simmer until potatoes are tender. Blend, add cream, season, and serve.

Flowers: Blooms You Can Eat

Edible flowers add a burst of color and unique flavors to culinary creations. However, caution is crucial, as not all flowers are safe for consumption. Accurate identification ensures a delightful and safe floral experience.

Identification Tips:

Petals and Sepals: Observe the number, arrangement, and color of petals and sepals.

Floral Structure: Note the structure of the flower, including the presence of pistils, stamens, and other reproductive parts.

Scent: Some edible flowers carry a distinct fragrance. Take note of any pleasant aromas.

Habitat: Consider the environment in which the flower is found. Certain species thrive in specific ecosystems.

Recipes:

1. Hibiscus Tea:

Ingredients: Fresh hibiscus flowers, hot water, honey.

Instructions: Steep hibiscus flowers in hot water. Sweeten with honey to taste.

2. Lavender Infused Syrup:

Ingredients: Fresh lavender flowers, sugar, water.

Instructions: Simmer lavender flowers in a sugar-water mixture until infused. Strain and use the syrup in beverages or desserts.

Seeds: Exploring Edible Seeds like Peas and Wheat

Edible seeds provide a rich source of nutrients and flavors. From the familiar peas to less common finds like wild wheat, identifying and harvesting these seeds opens up a world of culinary possibilities.

Identification Tips:

Seed Shape and Size: Examine the shape, size, and color of the seeds.

Seed Pod Characteristics: Note the structure of the seed pod if applicable.

Growing Patterns: Understand where and how the plant produces seeds. Some may be found in clusters, while others are singular.

Recipes:

1. Wild Rice Pilaf:

Ingredients: Wild rice, assorted foraged seeds, vegetables, broth.

Instructions: Cook wild rice, saute foraged seeds and vegetables, mix with cooked rice, add broth, and simmer until flavors meld.

2. Pea and Mint Hummus:

Ingredients: Fresh peas, chickpeas, garlic, lemon juice, mint leaves, olive oil.

Instructions: Blend cooked peas, chickpeas, garlic, lemon juice, and mint. Drizzle with olive oil and serve with crackers.

Roots: Harvesting Edible Roots, Including Onion and Garlic

Delving beneath the surface, edible roots offer a wealth of flavors, textures, and nutritional benefits. Identifying these hidden treasures requires careful consideration of both visual and olfactory cues.

Identification Tips:

Root Shape: Note the shape and size of the root, whether it's tuberous, bulbous, or fibrous.

Smell: Some edible roots, like garlic and onion, are easily identified by their distinctive aroma.

Soil Conditions: Understand the type of soil in which the plant thrives, as this can provide clues about the root's characteristics.

Recipes:

1. Roasted Wild Root Vegetables:

Ingredients: Assorted foraged roots (carrots, parsnips, wild tubers), olive oil, herbs.

Instructions: Toss cleaned and chopped roots with olive oil and herbs. Roast until tender.

2. Wild Onion and Garlic Soup:

Ingredients: Wild onions, wild garlic, potatoes, broth, cream.

Instructions: Saute chopped wild onions and garlic, add potatoes and broth, simmer until potatoes are soft. Blend, add cream, and season to taste.

Fruits: Sweet and Savory Wild Treats

Wild fruits offer a burst of natural sweetness and unique flavors. Identifying edible fruits involves understanding ripening stages, color changes, and growth patterns.

Identification Tips:

Color and Texture: Observe the color, texture, and shape of the fruit.

Ripeness Indicators: Learn the signs of ripeness, such as changes in color, softness, or aroma.

Plant Characteristics: Understand the overall growth habit and habitat of the fruit-bearing plant.

Recipes:

1. Wild Berry Jam:

Ingredients: Assorted wild berries, sugar, lemon juice.

Instructions: Cook berries with sugar and lemon juice until thickened. Pour into sterilized jars for homemade jam.

2. Elderflower Cordial:

Ingredients: Elderflowers, sugar, water, lemon.

Instructions: Steep elderflowers in a sugar-water mixture with lemon zest. Strain and use the cordial in beverages.

Stems: Edible Stalks and Shoots in Nature

Stems and shoots of certain plants contribute unique textures and flavors to culinary creations. Identifying edible stalks involves understanding growth patterns and differentiating between tender shoots and fibrous stems.

Chapter 12: Foraging Techniques

Foraging for edible wild plants is a delicate dance between exploration and preservation. Adopting ethical foraging practices, implementing sustainable harvesting tips, and honing the skill of locating and recognizing edible plant habitats are crucial aspects of becoming a conscientious forager In this comprehensive guide, we delve into the nuanced techniques that elevate foraging into a harmonious interaction with nature.

Ethical Foraging Practices

1. Leave No Trace:

Approach foraging with a "leave no trace" mindset Minimize your impact on the environment by avoiding unnecessary damage to plants, habitats, and wildlife.

2. Harvest in Moderation:

Practice moderation when harvesting Avoid depleting entire populations of a species in a single area Select only what you need, allowing the plant population to regenerate.

3. Respect Local Regulations:

Familiarize yourself with local regulations regarding foraging. Some areas may have restrictions or guidelines to protect delicate ecosystems or endangered species. Comply with these regulations to preserve biodiversity.

4. Cultivate Cultural Awareness:

Acknowledge and respect the cultural significance of certain plants in the region. Some plants may have historical or spiritual importance to local communities, and their sustainable use is vital for cultural preservation.

5. Avoid Rare or Threatened Species:

Refrain from foraging rare or threatened plant species. Focus on abundant and resilient plants to ensure the continued health of ecosystems.

6. Tread Lightly:

Be mindful of your surroundings. Stick to established trails when possible and avoid trampling on delicate vegetation. Respect the habitats of wildlife and nesting birds.

7. Educate Others:

Share your knowledge of ethical foraging practices with fellow foragers and nature enthusiasts. Promote responsible foraging to ensure a collective effort in preserving natural resources.

8. Consider Wildlife:

Remember that foraging impacts not only plants but also the wildlife dependent on them. Be conscious of nesting birds, insects, and other creatures in the area.

9. Be Selective in Harvesting:

Choose plants at their peak, ensuring the best flavor and nutritional value. Avoid harvesting plants that show signs of stress or disease.

10. Practice No-Digging:

When harvesting roots, use sustainable methods that don't involve uprooting the entire plant. Harvest only a portion of the roots, allowing the plant to regenerate.

Sustainable Harvesting Tips

1. Diversify Your Harvest:

Avoid over-harvesting a single species by diversifying your foraged finds. Harvesting a variety of plants reduces pressure on individual populations.

2. Know Your Impact:

Understand the life cycle of the plants you forage. Recognize how your harvesting practices may impact the plant's ability to reproduce and thrive.

3. Seasonal Awareness:

Be attuned to the seasons and the life cycles of different plants. Harvesting in sync with natural cycles allows for sustainable practices and ensures a continuous supply of edible plants.

4. Harvest Invasive Species:

Consider harvesting invasive plant species when appropriate. Removing invasive plants can contribute to the restoration of native ecosystems.

5. Encourage Natural Regeneration:

Allow plants to naturally regenerate by leaving some fruits, seeds, or roots behind during harvesting. This ensures the propagation of the species.

6. Adopt Permaculture Principles:

Integrate permaculture principles into your foraging practices. Mimic natural ecosystems, fostering biodiversity and resilience in the areas you explore.

7. Support Local Conservation Efforts:

Contribute to local conservation initiatives. Participate in or support projects aimed at preserving natural habitats and promoting sustainable foraging.

8. Learn from Indigenous Practices:

Indigenous communities often have time-tested sustainable harvesting methods. Learn from and respect their traditional knowledge, incorporating sustainable techniques into your foraging approach.

9. Consider Cultural Planting:

If allowed, consider cultivating certain edible wild plants in your garden. This not only supports sustainable harvesting but also provides an opportunity to learn about the plant's life cycle.

10. Monitor and Adjust:

Regularly assess the areas you forage. If you notice signs of over-harvesting or changes in plant populations, adjust your practices to promote sustainability.

Locating and Recognizing edible plant habitats

1. Understand Ecosystems:

Familiarize yourself with different ecosystems in your region. Learn about the types of soil, climate, and vegetation that characterize each habitat.

2. Explore Diverse Environments:

Foraging in various environments exposes you to a wider array of edible plants. Explore forests, meadows, wetlands, and coastal areas to discover diverse habitats.

3. Observation Skills:

Hone your observation skills. Pay attention to plant associations, noting which species tend to grow together in specific habitats.

4. Learn Indicator Plants:

Identify indicator plants that signal specific habitat types. For example, certain wildflowers or mosses can indicate soil acidity or moisture levels.

5. Study Plant Adaptations:

Understand how plants adapt to their environments. Some species may thrive in open, sunny areas, while others prefer shaded or damp conditions.

6. Seasonal Changes:

Recognize how plant habitats change with the seasons. A meadow that is lush with spring wildflowers may offer different foraging opportunities in late summer or fall.

7. Utilize Field Guides:

Consult field guides specific to your region. These guides often include information about common edible plants and their preferred habitats.

8. Join Local Foraging Groups:

Connect with local foraging groups or naturalist organizations. Participate in guided forays to learn about plant habitats from experienced foragers.

9. Take Soil Samples:

Understand the soil composition in different areas. Some plants may prefer acidic soils, while others thrive in alkaline or loamy soils.

10. Keep Records:

Maintain a foraging journal documenting the locations where you find different plants. Over time, this becomes a valuable resource for recognizing patterns and understanding plant distributions.

By integrating ethical foraging practices, sustainable harvesting tips, and a keen awareness of plant habitats into your foraging techniques, you not only enrich your foraging experiences but contribute to the preservation of ecosystems. Let your foraging journey be a harmonious dance with nature, fostering a deep connection with the land and its edible wonders.

Recipes and Culinary Tips

Cooking Techniques for Foraged Ingredients

Before delving into the recipes, let's explore some essential cooking techniques that elevate the use of foraged ingredients. These methods not only enhance flavors but also ensure a nuanced culinary experience:

1. Blanching and Shocking:

Preserve vibrant colors and nutrients in greens like nettles or dandelion by briefly blanching them in boiling water, then shocking in ice water.

2. Infusions and Tinctures:

Extract the essence of foraged herbs like mint, chamomile, or elderflowers through infusions or tinctures. Incorporate these into beverages, dressings, or desserts.

3. Dehydrating for Preservation:

Extend the shelf life of mushrooms or berries by dehydrating them. Rehydrate as needed for soups, stews, or snacks.

4. Fermentation for Flavor:

Immerse wild ingredients, such as ramps or garlic scapes, in fermentation for a distinctive tangy flavor. Ideal for pickles, relishes, or condiments.

5. Smoking Wild Game with Aromatic Woods:

Infuse wild game with unique flavors by smoking it using foraged aromatic woods like apple or cherry.

6. Wild Herb Compound Butters:

Elevate grilled meats or vegetables by crafting compound butters with foraged herbs like thyme, oregano, or sage.

7. Wild Flower Syrups:

Craft syrups from edible flowers like violets or elderflowers. Versatile for cocktails, desserts, or drizzling over pancakes.

8. Pickle Flower Buds:

Indulge in the tangy uniqueness of pickled flower buds, such as dandelion buds or nasturtium buds. Perfect for salads or appetizers.

9. Foraged Fruit Preserves:

Savor the seasonal flavors year-round by transforming foraged fruits into preserves or jams. Ideal with cheese or spread on toast.

10. Herb-infused Oils:

Enhance your culinary creations by infusing oils with foraged herbs like rosemary, thyme, or garlic chives. Perfect for cooking or as a finishing touch for salads.

11. Wild Mushroom Broth:

Create a robust broth using foraged mushrooms, a versatile base for soups, stews, or risottos.

12. Foraged Fruit Sorbet:

Embrace the refreshing taste of nature by blending foraged fruits with simple syrup, freezing, and churning into a delightful sorbet.

13. Wild Herb Rubs:

Elevate your meat dishes by creating herb rubs with dried foraged herbs. Perfect for grilling or roasting.

14. Acorn Flour Pancakes:

Infuse your breakfast with nutty goodness by substituting part of the regular flour in pancake batter with acorn flour.

15. Foraged Flower Infused Honey:

Transform honey into a delicacy by infusing it with edible flowers like lavender, chamomile, or elderflowers. Ideal for desserts or tea.

Recipes Incorporating Wild Plants into Everyday Meals

Now armed with these techniques, let's embark on a culinary adventure, infusing everyday meals with the flavors of foraged ingredients:

1. Wild Greens Omelette:

Ingredients: Foraged wild greens (dandelion, lamb's quarters), eggs, feta cheese, olive oil.

Instructions: Saute wild greens, mix with beaten eggs, cook until set, add feta, and fold into a delicious omelette.

2. Nettle and Potato Hash:

Ingredients: Foraged nettles, potatoes, onions, garlic, olive oil.

Instructions: Sauté diced potatoes, onions, and garlic. Add blanched nettles and cook until potatoes are golden brown.

3. Chickweed Pesto Pasta:

Ingredients: Foraged chickweed, basil, pine nuts, garlic, Parmesan cheese, olive oil, pasta.

Instructions: Blend chickweed, basil, pine nuts, garlic, and Parmesan. Toss with cooked pasta and drizzle with olive oil.

4. Wild Mushroom Risotto:

Ingredients: Foraged wild mushrooms, Arborio rice, onions, vegetable broth, white wine, Parmesan cheese.

Instructions: Sauté mushrooms and onions, add Arborio rice, deglaze with white wine, and slowly incorporate vegetable broth until creamy. Finish with Parmesan.

5. Dandelion Flower Fritters:

Ingredients: Dandelion flowers, batter (flour, water, salt), oil for frying.

Instructions: Dip dandelion flowers in batter, fry until golden brown. Serve as a unique and delightful appetizer.

6. Stuffed Grape Leaves with Purslane:

Ingredients: Foraged grape leaves, cooked rice, purslane, lemon, olive oil.

Instructions: Roll purslane and cooked rice in grape leaves. Drizzle with olive oil and lemon juice. Steam until tender.

7. Wild Garlic Mashed Potatoes:

Ingredients: Foraged wild garlic, potatoes, butter, cream, salt, pepper.

-Instructions: Boil potatoes, mash with butter and cream. Mix in finely chopped wild garlic, season to taste.

8. Rose Hip Jam:

Ingredients: Foraged rose hips, sugar, lemon juice.

Instructions: Simmer rose hips with sugar and lemon juice until thickened. Use as a sweet spread for toast or desserts.

9. Elderflower Lemonade:

Ingredients: Foraged elderflowers, lemons, sugar, water.

Instructions: Steep elderflowers in sugar water, add lemon juice, and dilute with water. Serve over ice for a refreshing beverage.

10. Wild Blueberry Pancakes:

Ingredients: Foraged wild blueberries, pancake mix, milk, eggs.

Instructions: Mix pancake batter, fold in wild blueberries, and cook until golden brown. Serve with maple syrup.

11. Sorrel and Potato Soup:

Ingredients: Foraged sorrel, potatoes, onions, vegetable broth, cream.

Instructions: Sauté onions and potatoes, add vegetable broth, simmer until tender. Blend, stir in chopped sorrel, and finish with cream.

12. Acorn Flour Banana Bread:

Ingredients: Foraged acorn flour, ripe bananas, flour, eggs, sugar.

Instructions: Incorporate acorn flour into banana bread batter for a nutty and unique twist on a classic.

13. Wild Berry Smoothie Bowl:

Ingredients: Assorted foraged berries, yogurt, granola, honey.

Instructions: Blend berries with yogurt, pour into a bowl, and top with granola and a drizzle of honey for a vibrant and nutritious breakfast.

14. Pine Needle-infused Vinegar:

Ingredients: Foraged pine needles, white vinegar.

Instructions: Steep pine needles in white vinegar for a few weeks. Strain and use the infused vinegar in salad dressings for a subtle piney flavor.

15. Chantilly and Thyme Butter:

Ingredients: Foraged chanterelle mushrooms, fresh thyme, butter.

Chapter 13: Common Edible Wild Plants

Foraging for wild plants not only connects us with nature but also offers a diverse array of flavors, textures, and nutritional benefits In this comprehensive guide, we explore detailed profiles and identification tips for common edible wild plants, delving into their nutritional value and culinary uses.

1. Dandelion (Taraxacum officinale)

Identification Tips:

Leaves: Distinctive toothed leaves forming a rosette.

Flowers: Bright yellow flowers with multiple petals.

Stem: Hollow, milky sap when broken.

Nutritional Value:

Rich in vitamins A and C.

Contains iron, calcium, and potassium.

Culinary Uses:

Salads: Fresh dandelion leaves add a peppery bite to salads.

Tea: Dried roots can be used to make a caffeine-free tea.

Pesto: Combine dandelion leaves with nuts, garlic, and Parmesan for a unique pesto.

2. Nettle (Urtica dioica)

Identification Tips:

Leaves: Toothed leaves with tiny hairs that sting when touched.

Stem: Square stem, often with a reddish tint.

Nutritional Value:

High in iron, magnesium, and vitamins A and C.

Contains protein and fiber.

Culinary Uses:

Soups: Blanch nettles to remove sting and add to soups.

Sauteed Greens: Cook like spinach and serve as a side dish.

Tea: Dried nettles make a nutritious herbal tea.

3. Chickweed (Stellaria media)

Identification Tips:

Leaves: Opposite, oval leaves with a single line of hairs.

Flowers: Small, white flowers with five deeply lobed petals.

Nutritional Value:

High in vitamins C and A.

Contains minerals like calcium and potassium.

Culinary Uses:

Salads: Fresh chickweed adds a mild, fresh flavor to salads.

Sandwiches: Use chickweed as a green addition to sandwiches.

Pesto: Blend with nuts, garlic, and cheese for a chickweed pesto.

4. Wild Garlic (Allium ursinum)

Identification Tips:

Leaves: Long, pointed leaves resembling lily of the valley.

Smell: Strong garlic aroma when crushed.

Nutritional Value:

Rich in allicin, a compound with potential health benefits.

Contains vitamins A and C.

Culinary Uses:

Pesto: Blend wild garlic with nuts, Parmesan, and olive oil.

Soups: Add chopped wild garlic to soups and stews.

Infused Oil: Make flavorful garlic-infused oil for cooking.

5. Sorrel (Rumex acetosa)

Identification Tips:

Leaves: Bright green, arrow-shaped leaves.

Taste: Tangy, lemony flavor.

Nutritional Value:

High in vitamin C and antioxidants.

Contains minerals like potassium and magnesium.

Culinary Uses:

Salads: Add fresh sorrel leaves to salads for a citrusy kick.

Sauces: Make a sorrel sauce for fish or poultry.

Soups: Incorporate sorrel into soups for a refreshing taste.

6. Wild Strawberries (Fragaria vesca)

Identification Tips:

Leaves: Toothed, trifoliate leaves.

Fruits: Small, red berries with seeds on the surface.

Nutritional Value:

Rich in vitamin C and antioxidants.

Contains fiber, manganese, and folate.

Culinary Uses:

Fresh Snacking: Eat them fresh for a sweet, juicy treat.

Desserts: Use in desserts like tarts, jams, or ice creams.

Infusions: Make strawberry-infused water or tea.

7. Wood Sorrel (Oxalis acetosella)

Identification Tips:

Leaves: Three heart-shaped leaflets.

Flowers: Five-petaled, yellow flowers.

Nutritional Value:

Contains vitamin C and oxalic acid.

Offers a tart, lemony flavor.

Culinary Uses:

Salads: Add wood sorrel leaves to salads for a zesty touch.

Garnish: Use as a garnish for soups or dishes.

Lemonade: Make a refreshing wood sorrel lemonade.

8. Wild Rose (Rosa spp.)

Identification Tips:

Leaves: Toothed, pinnate leaves.

Flowers: Distinctive five-petaled, fragrant blooms.

Nutritional Value:

Rich in vitamin C and antioxidants.

Petals are edible, while the hips are used for making tea or jams.

Culinary Uses:

Rose Petal Jam: Make jam from fragrant rose petals.

Tea: Use dried rose hips for a vitamin C-rich herbal tea.

Infused Honey: Infuse honey with rose petals for a floral sweetness.

9. Burdock (Arctium spp.)

Identification Tips:

Leaves: Large, heart-shaped leaves with woolly undersides.

Roots: Long, brown taproot.

Nutritional Value:

Rich in inulin, a prebiotic fiber.

Contains vitamins B6, potassium, and antioxidants.

Culinary Uses:

Root Vegetable: Roast or stir-fry burdock roots.

Tea: Use dried burdock roots to make a cleansing herbal tea.

Pickles: Make burdock pickles for a crunchy, tangy treat.

10. Violets (Viola spp.)

Identification Tips:

Leaves: Heart-shaped, slightly serrated leaves.

Flowers: Distinctive five-petaled, violet-colored blooms.

Nutritional Value:

Rich in vitamins A and C.

Edible flowers and leaves.

Culinary Uses:

Salads: Use fresh violet leaves and flowers in salads.

Candies: Make violet candies or syrups for desserts.

Infusions: Infuse violet flowers in water or vinegar.

11. Purslane (Portulaca oleracea)

Identification Tips:

Leaves: Succulent, paddle-shaped leaves.

Stems: Reddish stems.

Nutritional Value:

High in omega-3 fatty acids.

Rich in vitamins A, C, and E.

Culinary Uses:

Salads: Add purslane to salads for a crisp, lemony flavor.

Stir-Fries: Stir-fry purslane with other vegetables.

Pickles: Make purslane pickles for a tangy side dish.

12. Plantain (Plantago spp.)

Identification Tips:

Leaves: Broad, ribbed leaves in a basal rosette.

Flowers: Long, slender spikes with tiny, four-petaled flowers.

Nutritional Value:

Contains vitamins A and C.

Known for its anti-inflammatory properties.

Culinary Uses:

Salads: Use young plantain leaves in salads.

Tea: Make a soothing plantain tea.

Edible Seeds: Harvest and use the seeds in baking or as a crunchy topping.

13. Blackberries (Rubus fruticosus)

Identification Tips:

Leaves: Toothed, compound leaves with usually three to five leaflets.

Fruits: Dark purple to black, aggregate berries.

Nutritional Value:

High in vitamins C and K.

Rich in antioxidants and dietary fiber.

Chapter 14: Safety Precautions

A Comprehensive Guide to Responsible Foraging

Foraging for edible wild plants is a delightful journey into the heart of nature's bounty. However, it comes with inherent risks, including the potential presence of poisonous look-alikes and the varied sensitivities individuals may have. This comprehensive guide aims to provide detailed safety precautions to ensure a positive and secure foraging experience.

1. Poisonous Look-Alikes and How to Avoid Them:

A. Wild Carrot (Daucus carota) vs. Poison Hemlock (Conium maculatum):

Wild Carrot:

Identification:

Hairy stem with a distinct, anise-like aroma.

Single dark purple flower in the center of an umbrella-like cluster.

Edible taproot resembling a carrot.

Poison Hemlock:

Identification:

Smooth, purple-spotted stem with no hairs.

Small white flowers arranged in umbrella-shaped clusters.

Entire plant is highly toxic.

Precautions:

Education: Thoroughly research and understand the characteristics of each plant.

Visual Verification: Examine stem, leaves, and flowers for accurate identification.

Smell Test: Wild carrot emits a distinct, pleasant aroma when crushed; poison hemlock does not.

B. Morel Mushrooms (Morchella spp.) vs. False Morels:

Morels:

Identification:

Hollow stem attached at the base.

Distinctive honeycomb appearance on the cap.

Varied in color, often tan, yellow, or black.

False Morels:

Identification:

Solid stem, often wrinkled or irregular in shape.

Not attached at the base.

Some species are toxic and can cause serious harm.

Precautions:

Cap Structure: True morels have a pitted cap, whereas false morels may have irregular or lobed caps.

Stem Attachment: Check if the stem is attached at the base (morel) or not (false morel).

Cross-Reference: Use reliable field guides and seek expert guidance to cross-reference your findings.

C. Chanterelle Mushrooms (Cantharellus spp.) vs. Jack-O'-Lantern Mushrooms:

Chanterelles:

Identification:

Trumpet-shaped with gills running down the stem.

Often vibrant orange or yellow.

Pleasant, fruity aroma.

Jack-O'-Lanterns:

Identification:

Gills are bioluminescent (glow in the dark).

Often grow in clusters.

Entire mushroom is toxic.

Precautions:

Gill Characteristics: Chanterelles have true gills; jack-o'-lanterns have bioluminescent gills.

Growth Patterns: Chanterelles usually grow individually; jack-o'-lanterns may cluster.

Aroma Test: Chanterelles have a fruity aroma, while jack-o'-lanterns lack a distinct fragrance.

D. Edible Berries vs. Toxic Berries:

Edible Berries:

Precautions:

Clear identification using reliable field guides.

Consider the color, shape, and growth pattern.

Toxic Berries:

Precautions:

Differences in color, shape, or growth patterns.

Avoid consuming unknown berries.

E. Wild Garlic (Allium ursinum) vs. Autumn Crocus (Colchicum autumnale):

Wild Garlic:

Identification:

Smells strongly of garlic when crushed.

Tubular leaves resembling lily of the valley.

Edible and safe when properly identified.

Autumn Crocus:

Identification:

Toxic plant with no garlic smell.

Broader, non-tubular leaves.

Avoid ingestion as it is poisonous.

Precautions:

Scent Recognition: Crushing wild garlic releases a strong garlic aroma.

Leaf Characteristics: Tubular leaves are characteristic of wild garlic; broader leaves indicate autumn crocus.

Avoidance: If unsure, refrain from consuming any plant that doesn't meet identification criteria.

Precautions Summary for Poisonous Look-Alikes:

Education is Key: Arm yourself with detailed knowledge through field guides, workshops, or expert guidance.

Visual Inspection: Examine plants thoroughly, noting specific characteristics for accurate identification.

Smell Test: Certain edible plants, like wild carrot or garlic, emit distinctive aromas when crushed.

Cross-Reference: Utilize multiple sources for verification, including reputable field guides or foraging apps.

When in Doubt, Discard: If uncertain about the identity of a plant, refrain from consumption to avoid potential poisoning.

2. Allergies and Sensitivities in Foraging:

A. Pollen Allergies:

Precautions:

Timing: Be aware of flowering seasons and consider pollen levels.

Allergy Medications: Carry allergy medications if you are prone to seasonal allergies.

B. Contact Dermatitis:

Precautions:

Gloves: Wear gloves when handling plants to prevent skin irritation.

Long Sleeves and Pants: Cover exposed skin to minimize contact with irritants.

C. Common Allergenic Plants:

Precautions:

Identification: Learn to recognize common allergenic plants like poison ivy, oak, or sumac.

Avoidance: Steer clear of identified allergens to prevent skin reactions.

D. Sensitivity to Foraged Foods:

Precautions:

Patch Testing: Introduce new wild foods gradually and monitor for adverse reactions.

Start Small: Consume small amounts initially to gauge individual tolerance.

Consultation: Seek advice from healthcare professionals if uncertain about sensitivities.

E. Wild Mushrooms and Individual Reactions:

Precautions:

Identification: Learn about potential allergens in certain mushrooms.

Moderation: Consume small amounts of new mushrooms initially to detect individual reactions.

Consult a Professional: Seek guidance from mycologists or healthcare professionals about potential sensitivities.

Precautions Summary for Allergies and Sensitivities:

Self-Awareness: Be aware of personal allergies to plants, pollen, or specific foods.

Protective Measures: Wear appropriate gear, such as gloves and long sleeves, to minimize skin contact.

Patch Testing: Consume small amounts of new foods initially to monitor for any adverse reactions.

Consult a Professional: Seek advice from healthcare professionals or allergists if unsure about personal sensitivities.

Foraging for edible wild plants is an enriching experience when approached with knowledge, caution, and respect. Adhering to these safety precautions will help ensure that your foraging adventures are not only enjoyable but also safe and sustainable. As with any outdoor activity, responsible foraging contributes to the preservation of ecosystems and the well-being of foragers. Happy foraging!

Foraging Etiquette and Conservation

A Comprehensive Guide to Sustainable Practices

Foraging for edible wild plants is a dynamic and rewarding experience that connects individuals with the rich biodiversity of nature. However, to ensure the longevity of ecosystems and the well-being of both plants and wildlife, it is essential to embrace responsible foraging etiquette and conservation practices. This comprehensive guide delves deeply into the principles and strategies for fostering a harmonious relationship between foragers and the environment.

Respecting Nature and Wildlife:

1. Stay on Designated Trails:

Purpose:

Preventing Environmental Impact: Trails are designed to minimize damage to delicate ecosystems and prevent soil erosion.

Preserving Habitats:* By adhering to established trails, foragers avoid disrupting natural habitats and the fragile balance within.

Etiquette:

Trail Use: Stick to designated trails, avoiding shortcuts or creating new paths.

Footwear Consideration: Wear appropriate footwear to minimize soil disturbance.

2. Minimize Disturbance to Wildlife:

Purpose:

Wildlife Conservation: Reducing stress on wildlife populations allows them to maintain natural behaviors and thrive.

Preserving Habitats: By minimizing disturbance, foragers contribute to the preservation of undisturbed habitats.

Etiquette:

Observe from a Distance: Use binoculars and maintain a respectful distance to observe wildlife without causing disruption.

Avoid Disturbing Nests: Be cautious during breeding seasons to avoid disturbing nesting sites.

3. Leave No Trace:

Purpose:

Ecosystem Integrity: Leaving no trace ensures that ecosystems remain intact and unaltered by human presence.

Preventing Pollution: Eliminating litter prevents pollution and maintains the pristine quality of natural environments.

Etiquette:

Pack It Out: Carry out all waste, including food scraps, wrappers, and any non-biodegradable materials.

Waste Disposal: Dispose of waste responsibly, following local regulations.

4. Avoid Overharvesting:

Purpose:

Sustainability: Harvesting in moderation allows plant populations to regenerate, ensuring long-term availability.

Biodiversity: Leaving some plants untouched contributes to biodiversity and ecological resilience.

www.ingramcontent.com/pod-product-compliance
Lightning Source LLC
Chambersburg PA
CBHW070555010526
44118CB00012B/1322